# THE SCIENCE OF SUPERHEROES

# THE SCIENCE OF SUPERHEROES

**Lois H. Gresh**
**Robert Weinberg**

WILEY

John Wiley & Sons, Inc.

Published by John Wiley & Sons, Inc., Hoboken, New Jersey
Published simultaneously in Canada



**Library of Congress Cataloging-in-Publication Data**

Gresh, Lois H.
The science of superheroes / Lois H. Gresh, Robert Weinberg.
p. cm.
Includes bibliographical references and index.
ISBN 0-471-02460-0 (cloth)
ISBN 0-471-46882-7 (paper)
1. Comic books, strips, etc.—History and criticism. 2. Science. I. Weinberg, Robert. II. Title.

PN6714 .G74 2002
741.5'09—dc21

2002071323

Printed in the United States of America

10 9 8 7 6 5 4 3 2 1

As always, dedicated in loving memory to Big Daddy Sam, who will always be my Superhero. Special thanks to Dan Gresh, who told me about the 1993 "Death of Superman" plot. Many thanks to our agent, Lori Perkins, and to Stephen S. Power of John Wiley & Sons for letting us write this book.

—Lois H. Gresh

To Larry Charet, who opened the first full-time comic book shop in Chicago thirty years ago and who cofounded the Chicago Comicon in 1976. A pioneer in the comic book marketplace and good friend for three decades.

—Robert Weinberg

On the internet at:
www.sff.net/people/lgresh
and
www.robertweinberg.net

# Contents

Preface xi

A Word about the Law xix

Introduction **Men of Steel, Feathers of Fury**
**by Dean Koontz** **xxi**

Chapter 1 **More Powerful than a Speeding Locomotive: Superman** **1**
The Superman Legend Begins 1
What Makes Superman Super? 3
Alien Visitors 4
The Drake Equation 8
Rare Earth? 12
A Question of Gravity 15

Chapter 2 **Rays—Cosmic and Gamma: The Fantastic Four and the Incredible Hulk** **19**
Humble Beginnings 19
A Fantastic Foursome 21
Frankenstein's Monster—Marvel Style 22
The Perils of Technobabble 25
The GFP Hulk 29

Chapter 3 **The Dark Knight: Batman** **33**
A NonSuper Superhero 33
The Science of Batman 35
The Gotham City Earthquake 43

Chapter 4 **Under the Sea: Aquaman and Sub-Mariner** **47**
Undersea Heroes 47
Our Aquatic Ancestors? 50
Breathing Underwater 57
Pressure 58
Fluid Breathing 60
Talking to Fish 61

Chapter 5 **Along Came a Spider: Spider-Man** **65**
With Great Power 65
The Power of a Spider? 70
Clones, Clones, and More Clones 77

Chapter 6 **Green Lanterns and Black Holes: Magic, Science, and Two Green Lanterns** **83**
Wanted: An Unlimited Power Source 85
The Life and Death of Stars 86
The Origin of Black Holes 90
Yellow Light 96

Chapter 7 **Of Atoms, Ants, and Giants: Ant Man and the Atom** **99**
Ant Man 99
The Square Cubed Law 101
The Atom 104
The Atom Exploded 107

Chapter 8 **Fast, Fast, Fast: The Flash** **115**
Introducing the Flash 115
Problems with Logic 117
The Speed Barrier 125

Chapter 9 **Good, Evil, and Indifferent Mutants: The X-Men** **129**
A Victory Snatched from the Ashes 129
The Case for Evolution 133
The Truth about Creationism 137
Creating the X-Men 142

Chapter 10 **Mysteries in Space: Science Fiction Superheroes** **145**
Super Science without Super Heroes 145
The Secrets of Other Worlds, Exposed! 147
Doomsday on Earth 150
Across the Ages 153
The Grandfather Paradox 157

Chapter 11 **The Right Stuff: Donald Duck** **161**
The Real Deal 161
The Duck Man 162

Appendix A **Who Missed the Cut?** **167**

Appendix B **The Professionals Speak** **171**

Bibliography and Reading List 183

Acknowledgments 189

Index 191

# Preface

**Any book dealing** with superhero comics requires some slight knowledge of the history of comic books and their brothers-in-arms, the pulp magazines. Consider this brief preface the necessary prep work for a big exam. Or the instructions that airline attendants tell you every time you get on an airplane. Stuff that you don't need to read again and again, but background material available just the same. The proper grounding to make your flight through the rest of the book a lot easier.

Comic book superheroes, like jazz, potato chips, and hard-boiled detective fiction, are a uniquely American invention. Their roots are English and European, and they can be traced all the way back to the adventure fiction of Sir Walter Scott, the Gothic novels of Horace Walpole, the fantastic voyages of Jules Verne, and the science fiction inventions of H. G. Wells. Still, superheroes are most clearly defined by the American dream of the heroic individual. One man against the odds, whether it be the forces of nature, a corrupt government, or foreign invaders, comic book creations like Superman, Batman, Spider-Man, and the Incredible Hulk are as crisp a reflection of the American character as Uncle Sam—who, for the record, once starred in his own comic book series.

At the beginning of the twenty-first century, it's hard to imagine what life was like sixty or seventy years ago. The Depression gripped America and the rest of the world. Jobs were scarce, work was hard, and hours were long. It was a time before television. In the early

1930s, weekly radio adventures were still years away. Movies were something you saw on Saturday afternoons, if at all. Baseball was popular, but basketball and hockey were virtually unknown, and football was just starting to stretch its muscles. The main source of entertainment for both young and old was reading.

Libraries were free but offered limited reading choices. The best books often had many people waiting to borrow copies. And few novels were aimed at teens or young adults. For them, there were the pulps.

The pulps were inexpensive fiction magazines published from approximately 1900 until 1955. Prices ranged from five cents to fifty cents, with most costing a dime. There were pulps dealing with every type of fiction imaginable, from westerns to love to sports to mysteries to science fiction. "Pulp" fiction referred not to the style or type of story published in the magazines, but to the cheap wood-pulp paper used by the publications to keep costs low. The common bond in all pulp fiction was not violence, blood, or even danger. The stories were written to *entertain*, and for five decades and millions of readers, that's what they did.

In the 1930s, nearly a hundred different pulp magazines crammed the newsstands. Though derided as containing lowbrow formula fiction, pulps were read by millions of consumers every month. Most important to us, the pulp magazines of the 1920s and 1930s were the home of most science fiction and fantasy literature in America. And from these pulp roots came superhero comic books.

Comic strip adventures for children and adults had been running in newspaper serial format for decades. In the early 1930s, Tarzan and Buck Rogers emerged from the pulps. Tarzan, the creation of writer Edgar Rice Burroughs, first appeared in the novel *Tarzan of the Apes*, which was published in *The Argosy* pulp magazine in 1912. The story was so popular that it led to numerous sequels, all first published in pulps, as well as a series of movies and eventually a newspaper comic strip.

Buck Rogers was the main character of a short novel, *Armageddon 2419*, written by Philip Francis Nowlan. It was published in the August 1928 issue of *Amazing Stories*, the first all–science fiction pulp.

The cover for that issue, featuring a man hovering over the ground using a flying belt, was reprinted numerous times and came to define science fiction for most people as "that Buck Rogers stuff." Interestingly enough, the cover had nothing to do with Buck Rogers, as it illustrated another story in the same issue. A second Buck Rogers story, *The Airlords of Han*, followed in *Amazing Stories* in 1929.

The two Buck Rogers stories caught the eye of newspaperman Flint Dille. He contacted Francis Nowlan, the author of the Buck Rogers adventures, and asked if there was any way Nowlan could convert the stories into newspaper comic strips. Nowlan, working with artist Dick Calkins, did exactly that, and soon Buck Rogers was one of the most popular daily and Sunday comic strips in America.

By 1933, the tremendous popularity of comic strips convinced publishers to issue monthly reprint collections. These first comic books, titled *Famous Funnies*, *Popular Comics*, and *King Comics* (named after the distributor), were aimed at readers who wanted to preserve their favorite strips in book form. They also helped readers who missed episodes of the strip published in the past month. These reprint editions were the first monthly comics, but none of them included new material.

It wasn't until 1935 that an ambitious pulp writer, Major Malcolm Wheeler-Nicholson, devised the concept of publishing a comic book featuring all new stories and characters instead of the newspaper strip reprints. Wheeler-Nicholson titled his comic *New Fun Comics* and soon started a second magazine titled *New Comics*. Names changed quickly, and *New Fun Comics* became *More Fun Comics*, while *New Comics* turned into *New Adventure Comics*.

Unfortunately, the Major's concept of new comic stories didn't attract many readers, and Wheeler-Nicholson soon was badly in debt. His company survived only through an influx of needed cash from two new partners, Harry Donnenfeld and Jack Liebowitz. A third title, *Detective Comics*, the first comic book specializing in one subject, was started by the trio in March 1937.

Still, comic books featuring new characters didn't attract much attention from the reading public. Buyers were more interested in purchasing comics that featured familiar comic strip favorites like Dick Tracy, Little Orphan Annie, or Buck Rogers than unknown characters like Little Linda or the Radio-Squad. There just weren't any original comic book characters who compared with Tarzan or Flash Gordon. In early 1938, broke and disillusioned, Major Wheeler-Nicholson filed for bankruptcy.

Fortunately, Donnenfeld and Liebowitz continued to have faith in entirely new comics. Rather than give up, the two men bought the Major's interest in the firm and continued publishing. Their determination paid off with the appearance of *Action Comics* #1, dated June 1938. Featured in the issue was the first *Superman* story by Jerry Siegel and Joe Shuster. The cover for that landmark issue featured Superman lifting a car over his head. It promised readers something unique and entirely different from what had been published before.

The actual start of the Superman epic took place five years earlier. Two teenage friends in Cleveland in the early 1930s, Jerry Siegel and Joe Shuster, published a mimeographed science fiction fan magazine they called *Science Fiction*. In the third issue, dated January 1933, Siegel wrote a story entitled "The Reign of the Super-Man." Shuster did the art for the story, which featured a supergenius as the villain. It was the beginning of a partnership that made publishing history.

Over the next several years, Siegel and Shuster continued collaborating, blending story and art to tell science fiction adventures. Their goal was to sell a science fiction comic strip to the newspapers, much like *Buck Rogers*. During that time, their original concept of a "Superman" evolved from an evil genius to a crime-fighting hero with super powers. What started out as a fan magazine story slowly changed into a daily comic book strip.

Both the artist and the writer were fans of the pulp magazines of the era. Thus, it wasn't surprising that their Superman kept his identity secret by using an alter ego—much like those of such well-known pulp characters as the Shadow, the Spider, and the Whisperer. Siegel once stated in an interview that Clark Kent was named after popular screen actor Clark Gable. However, it seemed highly coincidental that the names of two of the most popular pulp heroes of the 1930s were *Clark* Savage Jr. (Doc Savage) and *Kent* Allard (The Shadow). And whereas Superman was nicknamed "the Man of Steel," Doc Savage was often described in his magazine as "the Man of Bronze."

The Superman comic strip was shown to all the newspaper comic publishers, but it never sold. However, when Donnenfeld and Liebowitz decided in 1938 to start a new comic book, *Action Comics*, Siegel and Shuster got their big chance. They were asked by the publishers for a Superman story for the first issue of the new comic.

Working under a tight deadline, the two men cut apart panels of their newspaper strip and pasted them together in comic book format to create the first Superman story. One panel from the story was blown up and used as the comic's cover. Supposedly, Donnenfeld balked at publishing a comic book showing a man lifting an automobile on the cover, but Liebowitz convinced him it would grab people's attention. Whether the story is legend or truth hardly matters. Liebowitz was right.

*Action Comics* #1 was a huge success. The story lifted new comic books out of debt and into business. *Action Comics* #1 sold out its entire print run of 200,000 copies and had magazine dealers demanding more.

Superman wasn't even featured again on the cover of *Action Comics* until issue #7. By then, the comic was selling over 500,000 copies a month. Donnenfeld and Liebowitz knew that the success of the comic was primarily due to the popularity of their lead character, Superman. In the summer of 1939, they released *Superman* #1, composed primarily of reprints of Superman stories from *Action Comics*. The book sold even better than *Action Comics*. Superhero comics were more than a success—they were a huge success. And in America, success breeds competition. Within a year, dozens of comic books featuring superheroes were competing with *Superman*. What later became known as "the Golden Age of comic books" had begun.

Not all comics published in the early 1940s featured superheroes. Familiar names like *Archie*, *Mickey Mouse*, and *Walt Disney's Comics and Stories* all had their beginnings in comics' Golden Age. But the fuel powering the engine that roared through American publishing was superheroes. There were dozens of characters with abilities ranging from super speed to the power to stretch and reshape their bodies into any imaginable size or object. All the world loved a superhero, whether it was Superman or Batman or the Flash or Plastic Man or Captain Marvel or the Human Torch or Submariner or the Blue Beetle or Ibis the Invincible.

However, all superheroes were not created equal. Whereas Superman was firmly rooted in science fiction, many of his contemporaries came from the opposite of the imaginative spectrum, fantasy. Captain Marvel, for example, was in reality newsboy Billy Batson, who turned into the super-powerful Captain Marvel when he uttered the magic word "Shazam." The Spectre, another character created by the writer of Superman, was the spirit of a murdered policeman, who returned from the afterlife to fight crime. Readers weren't particular about their heroes' origins or whether their powers were believable or not. The public wanted entertainment, and that's what they got.

For over sixty years, superhero comics have remained split into two categories: heroes and heroines powered by the wonders of science; and their counterparts, costumed crusaders gifted by magic. Often, the differences were slight, and supernatural and superhuman characters were more similar than they were different. It was mostly a matter of definition, as the dividing line between advanced science and ancient magic was barely noticeable. In the 1950s, science was seen as the solution to the world's problems and most superheroes were the children of advanced technology. In the 1970s, the counterculture revolution saw the rise of magical and supernatural champions. For the past two decades, the pendulum has swung back and forth from science to magic, from *Gen 13* and *Dr. Solar* to *Spawn* and *Sandman*. It's a rivalry that probably will continue for as long as comic books are published.

This book, obviously, deals with the comic book heroes of science fiction. While we have a great fondness for supernatural and magical characters, by their very nature they can't be explained by logic or scientific expertise. That's not the case with Superman, the Flash, and the Incredible Hulk, to name just a few. Coming from the world of science and technology, they can be put under a microscope and studied and analyzed. Their powers, origins, and adventures can be divided into what's possible, what might someday be possible, and what will never be possible. That's exactly what we plan to do in this book.

However, it's not going to be as easy as it sounds. That's because of a perception that began in the early days of comic books and remains with us today. It's the belief that stories told with pictures as well as words are only for children. That comic books aren't for adults.

From the 1940s through the 1960s, popular fiction in America went through major changes. The pulps gradually turned into paperbacks. Hardboiled mysteries, a staple of the pulps, became respectable. Science fiction, once considered that "crazy Buck Rogers stuff," gained in popularity and became a staple of paperback publishing, though it wasn't until the 1970s that the genre finally became accepted as literature. In general, as the readers of pulp magazines grew up, they took their tastes with them. What was considered cheap genre junk in 1940 became mainstream fiction in the 1950s. Except for comic books.

Because comics were considered by their publishers to be stories for children, real science was never a concern. The publishers felt no one would notice. They were correct because anyone who read comic books expecting actual science quickly abandoned them. While science fiction novels and short stories grew increasingly accurate in their depiction and use of science throughout the 1950s and 1960s, comic books went in the opposite direction and became less accurate. This trend in comics continues through the present day. Fans raised on pseudoscience and superheroes become writers and continue to write the same type of material as they once read.

Superheroes, nonetheless, offer us an ideal springboard into many realms of science. For the possible, as we'll see, is often much more fascinating than the impossible.

# A Word about the Law

**Before beginning our** odyssey of comic book science, we need to consider something scientists think obvious but which might baffle laypersons. It concerns perhaps the most important scientific discovery made in the past hundred years, and yet, most people don't know anything about it or how it affects our knowledge of our world and our entire universe. If you only learn one fact from this book, the answer to this question is the one worth remembering: *How can we assume that the laws of physics work the same throughout the universe as they do on Earth?*

Maybe, just maybe, in another galaxy, magic works and science doesn't. How can we know for sure? Or as more than one student has asked, *Just because things work a certain way here, how are we so sure they don't work differently somewhere else?* Or as expressed by many nonscientists when confronted by unbreakable laws, *How do we know that in a hundred years we won't be able to travel faster than the speed of light?* Is the real problem that scientists aren't willing to admit they could be wrong? Paraphrasing that thought, *With science constantly advancing, how can we say that anything's impossible?*

In a paper published in 1917 titled "Cosmological Considerations on the General Theory of Relativity," Albert Einstein stated what he called the cosmological principle. In Einstein's words: *No average property of the cosmic medium defines a preferred place or a preferred direction in space.* To put it in simpler terms, the universe obeys a certain set of laws that are the same everywhere. The cosmological principle implies that there is no center to the universe and that the universe is the same in all directions and the same when considered on the largest possible scale. To use more formal terminology, the universe is isotropic and homogeneous. Thus the laws of physics that are true on Earth are true everywhere else in the universe.

The word "theory" always bothers nonscientists. They argue that if something's a theory, then why do we assume it's a fact? In the past twenty years, we've heard this argument repeatedly, not as much relating to the general theory of relativity as to the theory of evolution. If it's a theory, argue creationists, then why do we teach it as truth in schools? The simple truth is that theories are concepts that have been examined and tested numerous times and never been proven wrong. Minor changes are sometimes required of theories, but the basic principles remain the same. When a theory has been proven so many times that it is no longer in doubt, then it's finally considered a law, such as the Three Laws of Thermodynamics.

The general theory of relativity is still listed as a theory because we can't fly all around the universe to check that it's correct from multiple locations. At least, we can't do that yet! Still, Einstein's theory has been shown to be true again and again in regard to every new discovery we make about other stars, other galaxies, quasars, and black holes.

So when we say, for example, that the speed of light is 300,000 km/sec everywhere in the universe, there is no *maybe* in that statement. It's true. Not because we say so. Not because scientists have closed minds. But because if it wasn't true, then the universe would make no sense and science would make no sense. The cosmological principle might be called a theory, but it's actually a law. And in science, laws aren't made to be broken.

**Introduction**

# Men of Steel, Feathers of Fury

## by Dean Koontz

**In my youth,** I was never interested in superhero comic books. I didn't dislike them, didn't swear an oath to track down and violently emulsify the writers and the artists who created them, didn't drive steamrollers through the printing plants that produced them, didn't turn lose packs of cats with weak bladders in stores that sold them, didn't purchase elaborate mail-order killing machines from Acme to eliminate the kids who read them, didn't build a time machine and travel into the past to waste the parents of the kids who read them and thereby ensure that those kids were never born. (By now perhaps you have begun to understand what I'm capable of doing when I really *do* dislike something.) I just didn't become excited by flying steel-bending, train-lifting, asteroid-deflecting, rhino-tossing, lava-walking guys with monumental muscles and flamboyant costumes. And Wonder Woman scared the crap out of me.

In my childhood—which many who know me would insist on defining as my first forty-three years, but which I define as my first ten—I was fascinated with certain *Little Lulu* comic books, which were filled with surrealism, and with Krazy Kat, for the same reason, but what *really* got my blood pumping and put a shine of mad glee in my eyes were comics featuring Uncle Scrooge McDuck and his easily infuriated nephew, Donald.

Scrooge often appeared in Donald's books, too, and they were always off to far-flung places, on great adventures that usually involved mysterious forces like Magica (the sorceress duck), Hassan Ben Jaild (the Arab bandit), ghosts, Mayan gods, and the phantom of Notre Duck. These brave and adventurous ducks had a strong sense of family, always looking out for one another, risking their lives for one another, which appealed to me because my father—a violent alcoholic, gambler, and womanizer—brought such disruption into my life that I had a far more fragile sense of family than I would have liked, and *no* sense of security.

Then there was Uncle Scrooge's money bin. This structure was a block-square, eight- or ten- or twelve-story monument to the glories of wealth that would make Bill Gates feel financially inadequate. Scrooge swam in this sea of coins and folding money as though it were water, and his love of money was portrayed not as a wicked glee, but as a pure, almost sweet delight. Scrooge didn't love money above all else. He would have given up his fortune to save his nephew and his grand nephew—Huey, Dewey, and Louie; in fact, if memory serves, he surrendered it more than once (not without a fight), but he always regained what was his. The infamous Beagle Boys, the aforementioned phantom of Notre Duck, giant robot robbers, and other nefarious types were never, in the end, a match for Uncle Scrooge, who was tough, smart, and clever—and who deserved his money because he'd earned it.

To understand the appeal Scrooge's money bin had for me when I was a kid, you have to know that we were poorer than those church mice everyone's always so full of compassion for, even though the annoying hymn-singing little rodents are *filthy vermin*. There wasn't any welfare in those days, and we never knew whether we would have money for next week's food. Even when my father worked, he frequently squandered his wages entirely on drink, card games, and women. Consequently, while other kids were reading about superheroes and were entertaining fabulous power fantasies, I was following Scrooge McDuck's adventures and entertaining *security* fantasies.

Superheroes were always out to save the world or at least one special corner of it. The ducks were always struggling to save their asses.

Superheroes were usually motivated by a selfless desire to do good and to fight Evil, with a capital E, often in the form of nasty archvillains. The ducks were usually fighting evil with a small e, desperate to keep what belonged to them or to get it back after it had been stolen. Superheroes were fearless, steely-eyed in the face of death. The prospect of death of even injury scared the poop out of the ducks, but when they had no choice but to fight, they fought with a squawking fury that Batman might envy. Superheroes seemed like demigods, figures of impossible fantasy, but to me the ducks seemd *real.*

By the time that I was eleven, blithely unaware that puberty was creeping toward me, but beginning to feel some strange changes taking place, I might have been expected to move on at last from ducks to superheroes, but instead I read science fiction novels. First the young-adult tales by Robert Heinlein, then adult science fiction by Heinlein, Ray Bradbury, Theodore Sturgeon, Jack Williamson, Hal Clement, and so many others: two books a week, sometimes three. Soon I came to expect at least a small degree of scientific veracity in my fantasy—and superheroes were contemptuous of scientific veracity.

Which brings us to *The Science of Superheroes* by Lois H. Gresh and Robert Weinberg. Although I eventually came to enjoy superheroes in movies—Superman, Batman, all their heroic brothers and sisters—I could never enjoy their adventures without mentally cataloguing the examples of nutball physics, biology, and botany with which their tales were replete. For that reason, I found this book to be a hoot from beginning to end. Ms. Gresh and Mr. Weinberg must have spent some time in institutions for the deranged, because well-balanced minds could not have conceived of this project. But thank God for their derangement, for they have produced a package of pure fun from first page to last. If, like me, you admire superheroes from a distance, or if you are a hardcore fan of them, you will enjoy this book as surely as you would enjoy waking one morning to discover that you are invincible, able to fly, and in possession of a totally cool costume behind which to hide your true identity.

## Chapter 1

# More Powerful than a Speeding Locomotive

## Superman

### The Superman Legend Begins

**The success of** Superman in *Action Comics* #1 propelled comic books from minor amusements to a mainstay of American entertainment. A small business franchise that seemed doomed to failure suddenly blossomed into a million-dollar industry. In 1938, no one ever heard of Superman. By 1940, he was the star of two comic book magazines selling over a million copies a month and had become an American icon.

Superman's surprise success story was comparable to only one other imaginary superstar's rise to fame—Mickey Mouse. Not surprisingly, it wasn't long before the Man of Steel ventured onto the big screen. The 1940s saw a series of visually stunning Superman cartoons from the Max Fleischer studios. Superman toys followed, as did games, and more comics featuring Superman stories like *World's Finest Comics*.

In 1949, aiming to expand their Superman franchise even further, DC Comics began another spin-off comic book series, *Superboy*, featuring the adventures of Superman as a teenager. Superboy was also featured in stories in *Adventure Comics*. *The Adventures of Superman* TV show, starring George Reeves, became one of the staples of 1950s television. The market continued to expand with Superman games, Superman clothing, and Superman giveaway toys in cereal boxes.

In more recent times, there have been Superman movies, several TV shows, and a new cartoon series. The "Death of Superman" storyline garnered international news coverage. A half-dozen different Superman comics crowd the shelves of comic book stores every month. It's estimated that Superman is one of the best-known fictional characters in the world, rivaled only by Sherlock Holmes—leading us to wonder why. What makes Superman so incredibly popular?

When Superman debuted in 1938, the only widely read science fiction stories were the adventures of characters like Flash Gordon and Buck Rogers, published in newspaper comic strips. Science fiction magazines featuring short stories and novels were a small part of the huge pulp market. It was a genre largely ignored by critics, who considered it junk literature aimed at teenagers. The few authors—like H. G. Wells—whose work was considered worth reading were called social visionaries, not science fiction writers.

Superman changed all that. Though comic books were supposedly aimed at children, everyone read Superman. Some issues of his comic sold over a million copies, with most copies circulating among several people. In a country with a population of forty million citizens, Superman became a national phenomenon. What made Superman so unique? Obviously, it was his *super* powers.

Originally, Superman's powers were super merely when compared to those of a normal human being. Over the years, as the comic evolved, as Superman faced greater challenges, the extent and nature of his powers increased. A Superman who could withstand the force of an artillery shell seemed inadequate when measured against the power of an atomic bomb. Thus his writers made Superman more and more invincible. By the 1960s, Superman had become so powerful that finding menaces for him to defeat became difficult. Stories lacked suspense because it was obvious that no power in the universe could defeat the "Man of Steel." More adventures relied on magic, which needed no logical explanation, or bizarre plot devices that caused Superman's powers to decrease.

Finally, in the mid-1980s, Superman's history was rewritten and his powers downgraded so that he no longer was invincible and invulnerable. His origin was retold, and he began again with a new history and background. Superman became less super and more man. He became a hero for the 1990s and beyond, a sensitive, New Age Man of Steel.

## What Makes Superman Super?

When we examine Superman, we need to remember that, in a sense, we're examining all the superheroes who follow. Superheroes have always been created with broad brushstrokes. Not a lot of time was spent on deducing the limits or nonlimits of our super characters. Even less attention was paid to their interaction with ordinary people and objects. When Superman lifts a car over his head to shake criminals to the ground, no one ever questions why the car doesn't fall to pieces. Nobody questions how Superman stays perfectly balanced on Earth while waving over his head an item that has a mass twenty times greater than his own.

How often have we seen Superman fly down and pull a car up by the roof into the sky? In the real world, there are few vehicles that would even hold together if Superman yanked them up by the roof. The car would probably continue forward, with the roof ripped off and held by Superman. Every time Superman lifts a building into the air, why don't all the bricks, held together by cement and pressure, suddenly start falling apart? Those are the types of ordinary problems that seem never to occur in any superhero adventures. Basically, superheroes perform super acts and the logic squad cleans up afterwards.

In Superman's first appearance in the 1938 *Action Comics*, we're informed "that he could leap one-eighth of a mile; hurdle a twenty-story building . . . raise tremendous weights . . . run faster than an express train . . . and that nothing less than a bursting shell could penetrate his skin!"[1]

Siegel and Shuster's explanation of Superman's powers, as given in *Action Comics* #1, left much to the imagination. Their main premise

---

[1] *Action Comics* #1, June 1938.

was that Superman came from a civilization much more advanced than ours and thus the inhabitants were physically more advanced than humans. By extrapolation, this argument implies that modern man is physically much stronger than Cro-Magnon man or Neanderthal man. Of course, our ancestors lived only a few hundred centuries before us, while Superman's race was described as being millions of years ahead of ours. A full page illustration in *Superman* #1 (Summer 1939) gave a "scientific explanation of Superman's amazing strength."

"Superman came to Earth from the planet Krypton, whose inhabitants had evolved, after millions of years, to physical perfection. The smaller size of our planet, with its slighter gravity pull, assists Superman's tremendous muscles in the performances of miraculous feats of strength!"[2]

Thus, Siegel and Shuster gave two explanations for Superman's extraordinary powers. He was an alien from a planet not in our Solar System, and the weak gravity of Earth compared to the gravity of his home world of Krypton gave him amazing strength. Both concepts came right from the pages of science fiction magazines of the time, and few readers questioned the logic of either assumption. However, we've made huge strides in science during the past six decades. Let's examine each premise based on today's knowledge.

## Alien Visitors

Alien visitors from outer space have been a basic staple of science fiction since H. G. Wells first wrote *The War of the Worlds* in 1895. It's one of the few science fiction concepts that has crossed into mainstream belief. In 1939, the same year that Superman gained his own comic book, thousands of people were convinced by an Orson Welles radio dramatization of Wells's novel that Martians had landed in New Jersey. Siegel and Shuster weren't writing anything outrageous in terms of what people thought to be true. Or still believe to be true. The only problem is dealing with cold scientific reality.

---

[2] *Superman Comics*, Summer 1939.

Do aliens exist? Are they intelligent? If so, have they visited us? That's the basic, underlying concept of Superman. These questions have spurred some of the hottest debates in modern astronomy. Let's examine each side of the arguments and see what they tell us about Superman.

*Pluralism*, the belief that the universe is filled with planets harboring intelligent life, has been examined by philosophers and scientists for thousands of years. It was first championed by the Greek atomist philosophers Leucippus, Democritus of Abedera, and Epicurus in the fifth century B.C. They believed that the Earth was produced by a chance collision of indestructible particles known as atoms. Since one world had been formed in such a fashion, they argued that other worlds with intelligent life were possible as well. Opposing the atomists were Aristotle and Plato, who argued that the Earth was unique and that no other worlds or intelligent life forms existed.

Needless to say, the Aristotelian view of the universe was accepted by the Catholic Church because that viewpoint placed man in a special place in the universe. However, in the late twelfth century, a number of scholars raised some serious religious arguments against Aristotle's belief that there was only one possible cosmos. Since God was omnipotent, these men declared, then stating God created only one universe was in a sense placing restrictions on God's power, which would thus imply that God was not all-powerful. In 1277, the church eased its stance on the unique nature of the universe. Catholic doctrine was revised to say that God *could* have created other worlds with intelligent beings, but didn't.

It was a very small step for science, but a major one. Following the same line of reasoning, Nicholas of Cusa in 1440 declared that whatever God could do would be done, a belief that became known as *plenitude*. Less than a century later, Copernicus argued convincingly that the sun was the center of the solar system and the Earth was merely a planet revolving about it. Copernicus wisely didn't delve into the theological ramifications of his discovery, but other scholars and philosophers were soon debating the possibility of life on other worlds.

For the next several hundred years, proponents and opponents of plenitude and pluralism argued about God's purpose in creating a universe filled with stars and planets. As usual in debates where men tried to explain God's purpose, neither side convinced the other that they were correct. Fortunately, by the mid-nineteenth century, developments in science and astronomy made such debates moot. God's intent faded in the face of the theory of evolution, and a scientific view of the universe slowly but surely replaced the religious one.

Still, while pluralism and plenitude were interesting theories, there was no factual evidence to back up either philosophy. Telescopes could show only so much. There were no canals on Mars, and the clouds of Venus didn't shroud gigantic oceans or primeval forests. The only aliens from other worlds appeared in science fiction books and magazines, or in comic books like *Superman*.

After years of stories about life on other planets, people started wondering where the aliens were and whether it was possible that we were the most intelligent species in the universe. If the galaxy was so huge and full of life, why hadn't other life forms contacted us?[3]

As always when a question is raised, someone was there with an answer. It's not surprising that the average citizen living in a Cold War atmosphere of distrust and misinformation was more than willing to believe that our government was concealing the truth about aliens from us.

In 1947, we suddenly "learned" from several "nonfiction" books and magazine articles that other eyes were watching. It was the beginning of the flying saucer craze.

Flying saucers were featured in innumerable magazine stories and tell-all books, and they dominated late-night talk radio. Saucers have remained in our skies for the past half century despite the lack of any conclusive evidence proving their existence. In 2001, surveys indicated that a majority of people in this country believe Earth has been visited by aliens.

Suddenly, the question wasn't whether aliens existed on other planets. Instead, the question became: Why were the aliens spying on us?

Flying saucers were a major setback for scientists trying to prove that extraterrestrial life existed in the galaxy. Frank Drake, one of the

---

[3]Of course, flying saucer advocates claim that aliens have contacted us. For purposes of this discussion, we'll consider only verified alien landings.

leading astronomers of the twentieth century, put it best when he stated:

> The problem is that no civilization can thrive on falsehood. In the end, false "knowledge" leads to wrong decisions, wrong choices of technologies, a wrong distribution of resources, wrong priorities, wrong choice of leaders. And civilization as a whole is the loser. A prime illustration of this is the distribution of resources invested in attempts to understand life in the universe. There is widespread public confusion as to the relative promise of pseudo-scientific studies of UFOs . . . as compared with true scientific programs to find life on other planets. . . . The consequence is that far more attention is given to the pseudoscience than to the real science.[4]

So whether you believe in flying saucers or think that they're an ongoing money-making hoax, where are the aliens? If they're here, why are they so shy? After all, building a ship that can navigate the far reaches of outer space takes a fairly sophisticated and advanced civilization—one a good deal more advanced than ours. They can't all be tongue-tied or hiding under sofa cushions. Surely one of them has something to say to the world at large. Even a quick "Hi" would satisfy most people.

That's a moral issue that Siegel and Shuster (and the many writers and artists who followed) never confronted in *Superman*. Clark Kent was raised in secret, so as not to reveal that he originally came from another planet. Only in the world of comic books would such a concept be accepted as a good idea. Imagine the worldwide excitement there would be on Earth if a flying saucer landed on the White House lawn and a humanoid figure stepped out to bring greetings from another planet.[5] A visit like that would change the world overnight. We'd actually know that we aren't alone in the universe; not to mention that we aren't the smartest or even the strongest kids in the neighborhood. But in comic books, all it meant was a new cop on the block.

---

[4]*Sharing the Universe*, 1998. Foreword, pp. i–ii.

[5]See *The Day the Earth Stood Still*, one of the best science fiction movies ever produced about this topic. It was filmed during a time when SF films weren't used just to sell toys and feature great special effects.

Which brings us right back to Superman's initial premise. Pluralism and plenitude are interesting theories but have no basis in fact. Are there alien civilizations in the galaxy? Do we have any proof at all that we are not alone in the universe other than religious doctrine?

## The Drake Equation

In the 1950s, astronomer Frank Drake proposed an equation to estimate the number of intelligent species in our galaxy, the Milky Way. This equation served as the rallying point for the earliest efforts to use radio telescopes to detect signals sent by other highly advanced civilizations. Run for months by Drake, Project Ozma had no success in detecting the all-important radio signals from other star systems. However, a far greater effort was organized by scientists and continues through this day. The Search for Extraterrestrial Intelligence (SETI) served as the background for Carl Sagan's book (later made into the movie) *Contact*.

The Drake Equation is a fairly simple multiplication problem.

$$N = R^* \times f_p \times ne \times f_t \times f_i \times f_c \times L$$

where

N is the number of intelligent civilizations in the galaxy (the number we are looking for).

$R^*$ is the birth rate of suitable stars for life in the Milky Way galaxy measured in stars per year.

$f_p$ is the fraction of stars with planets.

ne is the number of planets in a star's habitable zone (which we define below).

$f_t$ is the percentage of civilizations that have the technology and desire to communicate with other worlds.

$f_i$ is the fraction of habitable planets where life does arise.

$f_c$ is the fraction of planets inhabited by intelligent beings.

L is the average in years of how long the technologically advanced civilizations last. In other words, how long is it from the time aliens invent radio to when their civilization either destroys itself or disappears?

The only phrase that may seem confusing is a "star's habitable zone." In simplest terms, the phrase refers to the imaginary shell around a star where the surface temperature of a planet in that shell would be conducive to the origin and development of life. As far as humanity is concerned, the habitable zone around a star is the space where planets exist that have water in liquid form, the most basic necessity for life. In our solar system, Earth is obviously in the habitable zone. Venus, which is too close to the sun, is not. Nor is Mars, which is too far away.

In the 1950s, when Frank Drake invented the Drake Equation, many of the numbers and fractions were not known. As our knowledge of astronomy grew, more of the numbers became available. Still, some were based more on hopes and beliefs than actual information.

A very popular theory about the universe believed by Carl Sagan and other space scientists is known as the Principle of Mediocrity (sometimes called the Copernican Principle). This theory, based entirely on logic, states that since Earth appears to be a quite typical and common planet, intelligence has a very high probability of emerging on any planet similar to Earth after 3.5 billion years of evolution.

In simplest terms, the Principle of Mediocrity states that Earth isn't special, so there should be lots of other planets with life on them.

Belief in the Principle of Mediocrity fuels the scientists who believe in SETI. It's also what makes the Drake Equation work. Without it, we'd most likely not believe stories about Superman or visitors in flying saucers. However, in the past decade, a growing number of scientists have been studying the Principle of Mediocrity, and they find it wanting. We'll discuss this idea in our next section.

For the moment, let's plug some numbers into the Drake equation. To put things in perspective, let's use the numbers that Carl Sagan and Drake himself used to see how many intelligent extraterrestrial civilizations are out there.

R* has been estimated by astronomers to be between 1 and 10 stars per year. Drake picked 5 as an average.

For $f_p$, the number of stars with planets, Sagan believed that a majority of stars had planets. In the past few years, we've actually located some. Let's be somewhat conservative and pick twenty percent, or one out of every five stars.

For ne, the number of planets that exist in the habitable zone, if we use our solar system as a model (the Principle of Mediocrity), then the number is one, since Earth is the only planet in our solar system we know for sure has water in liquid form.

For $f_l$, the percentage of worlds like Earth where life begins, Drake and Sagan chose one hundred percent, again using Earth as their model.

For $f_c$, the percentage of planets with intelligent life, SETI scientists argue that evolution over billions of years leads to intelligence, so again the percentage could also be one hundred percent.

For $f_t$, intelligent species who develop the technology and the desire to communicate with other worlds, Drake estimates that this value is one hundred percent.

The math is pretty basic. Multiply all the numbers we have so far and (surprise, surprise) we end up with an equation that $N = L$. This is the same result that Frank Drake and Carl Sagan arrived at years ago: the number of intelligent civilizations in the galaxy equals the average lifetime of technologically advanced civilizations.

Again, let's assume that Earth is average (using the Principle of Mediocrity). If our civilization self-destructed next year due to nuclear war or the release of a deadly plague virus,[6] then L would be approximately 100, meaning that our galaxy would be home to one hundred alien civilizations.

Considering that there are somewhere between 200 and 400 billion stars in our galaxy, we suddenly are faced with the possibility of one civilization per two to four billion stars. It's no wonder we haven't been contacted by aliens. Reducing it to more human terms, it would be as if two people were born on the Earth during the past fifty years, separated not only by time but also by thousands of miles. Neither has any clue about the other except that they have the same birthmarks. Then somehow, they must find each other, searching on foot.

Frank Drake and Carl Sagan both knew that L, the lifetime of a technologically advanced civilization, was the great stumbling block in the Drake equation. However, both men were not only scientists but

[6]Forty years ago, we'd only consider nuclear holocaust as our possible doom, but now a manmade plague distributed by terrorists seems just as likely.

optimists. Drake felt that a technological civilization might last for ten thousand years. Thus, he estimated that there were 10,000 advanced civilizations in the galaxy.[7] This would leave us with one civilization per twenty to forty million stars, still somewhat of a daunting search. Other scientists believe that number to be much too low. They estimate that there could be hundreds of thousands of such civilizations. Which would mean we're not as alone as we thought.

Most important to our concerns, the Drake equation, working with the figures cited, gives us estimates ranging from 100 to 10,000 civilizations in the galaxy. All we need for Superman is one.

In *Superman* #132, it's stated that Krypton was three million light years from Earth. That's a pretty amazing distance, even for comic books. Our home galaxy, the Milky Way, for comparison, is only 100,000 light years across. There are approximately twenty other galaxies, some much larger than the Milky Way and some much smaller, within three million light years, increasing our range of possible civilizations from 2,000 to 200,000.

We'll resist the temptation to speculate on how Jor-El (Superman's father) and his fellow Kryptonian scientists located modern-day Earth or sent a rocket to it. Instead, we'll just assume that their science is a million years ahead of ours. What matters is that we've gone from a galactic scale to a universal one. Even if there are only a few hundred civilizations per galaxy, there are a lot of galaxies in the known universe. A recent estimate placed the number at fifty billion galaxies. Assuming one hundred civilizations per galaxy, that still results in five thousand billion (5,000,000,000,000) intelligent civilizations in the known universe. If we take Frank Drake's more optimistic guess, we're talking about 500,000,000,000,000 (five hundred trillion!) advanced civilizations in the universe.

Suddenly, the existence of Krypton seems a lot more possible. But don't tell Peter D. Ward and Donald Brownlee that. Because they've raised some serious doubts about the Drake equation.

---

[7]Carl Sagan, who was even more optimistic than Drake, estimated in 1974 that there might be a million civilized planets in our galaxy.

## Rare Earth?

In their 2000 book, *Rare Earth: Why Complex Life Is Uncommon in the Universe*, Peter F. Ward and Donald Brownlee discuss the Principle of Mediocrity. The two scientists argue that perhaps we've gone too far in trying to prove that man isn't special. They argue that in our attempts to understand the universe surrounding us, we've downsized the significance of life on Earth. They propose that perhaps life is not common and the Principle of Mediocrity isn't true. Maybe, as the ancient religious thinkers believed, mankind is unique.

It's a startling proposition, but the two men build a compelling case. Chapter by chapter, they examine each factor in Frank Drake's famous equation and arrive at totally different conclusions.

The basic problem with the Drake Equation is that it's a series of numbers multiplied together to give us a final answer. In any multiplication problem, if any one number is zero, the answer is zero. If any one number is a very small fraction, the answer becomes a small fraction. If the assumptions used to produce those numbers are incorrect, then the numbers are invalid. In the Drake Equation, too many of the numbers are based entirely on speculation, hope, and faith, not on fact.

Let's examine four of the most troubling figures in the Drake Equation. Instead of taking the optimistic viewpoint adopted by the people working on SETI, let's instead look at them through the much more pessimistic lens of *Rare Earth*.

For example, $f_p$ is the fraction of stars with planets. Our solar system has nine planets. Carl Sagan argued that an average solar system most likely would have ten or more. Other noted scientists of the 1970s and 1980s felt that ten planets was a good estimate. However, major strides in astronomy during the past decades have caused astronomers to rethink this belief.

In the last ten years, scientists have discovered twenty-seven planetary bodies circling other stars. All of the planets we've located are huge, about the size of Jupiter, the largest planet in our solar system. Astronomers studied numerous stars to find the twenty-seven stars with huge planets circling them. There's no method yet developed to locate smaller, Earth-sized planets. So, the guess that $f_p$ is one of every five stars with planets could be extremely high. We probably won't have a good estimate on the average until we start traveling to other solar systems.

Ne is the number of planets in a star's habitable zone. Until recently, the habitable zone has been defined as the appropriate distance from a star that enables liquid water to exist and complex life to develop. Earth is the only planet in the habitable zone of our solar system. It's possible that some sort of simple biological life might exist or once have existed on Mars, but that's not been proven.

Ward and Brownlee argue in their book that based on what we've learned about astronomy in the past few decades, it's clear that habitable zones are a lot more complicated than anything imagined by Drake and Sagan in the 1960s or 1970s. For example, they point out that the presence of Jupiter, a massive gas giant much farther out in our solar system, was a critical factor in life developing on Earth. Jupiter's immense gravitational pull attracted most comets to it instead of allowing them to crash into Earth. Without a Jupiter-sized planet serving as this type of shield against stray comets, life on Earth would have been subject to mass extinction events and planetary disasters caused by space collisions.

Therefore, the habitable zone of a solar system isn't merely based on the location of a planet in a solar system, but the location of other planets in the system, as well. Which makes the existence of habitable zones a great deal less probable than once considered.

Ward and Brownlee take habitable zones a step further by considering the zone of space where animal life, not merely biological life, could develop. Biological life, such as primitive bacteria, can exist in extreme heat or extreme cold. Humans can't, and the difference needs to be taken into account.

$F_i$ is the fraction of habitable planets where life does arise. In *Rare Earth*, the authors examine the length of time, measured in billions of years, necessary for a habitable zone to exist in relative stability for evolution to take place. Using Earth as our model, that time zone needs to be at least three billion years long. They point out that our sun, a G2 type star, has a lifetime of ten billion years, more than enough time for complex life to develop.

However, G2 suns are not the most common stars in the galaxy. That honor belongs to M stars, which have a mass of only about 10 percent of our sun. As these stars don't emit nearly as much heat as Sol, the habitable zone around them is much closer to the star itself. Planets would need to orbit much nearer to the sun, which leads to a host of problems. Gravitational tidal effects from the star lock the planet into an orbit where only one side of the planet faces the sun—an orbit like that of Mercury. It's an orbit that's not conducive to human life.

Going in the other direction, stars more massive than our sun have a much shorter lifetime. Sol is predicted to remain stable for ten billion years. A star fifty percent more massive than our sun would last only two billion years before entering the red giant stage. When a regular star transforms into a red giant, all planets in the original habitable zone in space are burned away, as new habitable zones are established millions of miles further out.

Big, hot stars like Sirius also generate a lot of their energy as ultra-violet light. UV light is fatal to biological molecules, so any star system with a high density sun wouldn't be the home of carbon-based beings. So $f_i$ might be a complex problem, involving habitable zones, the structure of the solar system, and the type of Suns at the center of the same system.

The situation becomes murkier and definitely not better. $f_i$ is the fraction of planets on which life develops intelligence. Frank Drake felt that every place where life began, intelligent life would arise. That's an optimistic viewpoint, based entirely on the fact that intelligent life developed on Earth. The more we learn about the slow, complex path of evolution from single-celled organisms to a walking, talking, thinking man, the less sure that number becomes. Drake and Sagan argued that intelligence was inevitable on any planet where life began. Many scientists now believe that considering the more than three billion years it took for complex, intelligent life to evolve on Earth, we were very lucky.

If we could compress time so that one second equaled ten thousand years, all of humanity's recorded history could be squeezed into one second. Mankind's rise from simple predator to ruler of the Earth fills three seconds. However, the time it took for one-celled organisms to evolve into intelligent life spans two and a half days. During that long, slow rise, paleontologists know of at least ten extinction-level events where more than half of all known life on Earth was destroyed. Optimists would argue that the development of intelligent life on Earth despite these ten extinction-level events demonstrates that complex life is inevitable. Pessimists would argue that we've been fortunate and the next extinction-level event could be our last one.

If $f_i$ is less than one hundred percent, then what of $f_t$, the number of alien races that will try to communicate with other species from another planet? Frank Drake and the scientists of SETI believe that percentage to be one hundred percent—that every race of beings in the universe wants to discover intelligent life elsewhere. They base their assumption on *our* behavior. But, aliens, being alien, probably will have little or nothing in common with us. They might not be curious, or a vast number of them might not want to use their resources to contact other races in space. Instead, they may spend their money on the poor, the homeless, and the hungry. $f_t$ could be one hundred percent, but it just as likely could be one percent. If the Drake Equation is to have any relevance, we need to consider both possibilities.

If we plug in all the worst-case scenario numbers into the Drake Equation, then estimate that L is 1,000 rather than 10,000, the result is that there may be only one civilization in the Milky Way galaxy. Like it or not, we may actually be alone in our galaxy.

This fact explains why Jor-El sent the last survivor of Krypton to Earth. Maybe it wasn't because he thought Earth needed a hero. Maybe it wasn't because Earth was the one planet where his son, Kal-El, the being who became known as Superman, would have the greatest chance of survival. Maybe Superman's father had no choice. Maybe there was no place else to go.

## A Question of Gravity

We've answered one of the major questions raised in *Superman* comics. Could there be other races in the cosmos? The answer is yes, there *could* be, though we're not sure how many or how close they might be to Earth. Their location isn't important, nor how they would send a rocket across millions of light years. Space travel isn't one of the themes of Superman. We're only interested in whether an alien could come to our world.

Let's assume Superman could indeed come to Earth. What powers would he possess that would make him a superman when compared to humans?

Going back to the original Superman of *Action Comics* #1, it's clear he has tremendous strength and can jump great distances, but he never flies. In *Superman* #4, for example, he runs from Metropolis to Oklahoma. Siegel and Shuster created a character they thought was believable based on the science of the time. There was no explanation for flight, so the best Superman could do was jump.

As the years passed and competition increased, Superman's powers grew as his creators continued to change the character to meet the demands of an ever-increasing audience. By 1943, Superman could fly at speeds faster than light (another impossibility). Needless to say, as his powers grew more incredible, so did his strength. In early issues of *Action Stories*, Superman lifts an automobile over his head. Within a few years, he's carrying buses packed with astonished riders. After a few more years, he's carrying ocean liners. By the 1960s, he's moving planets.

Siegel and Shuster's original comic book concept was that Superman's tremendous strength was the result of being born on a high-gravity planet. Earth's gravity was much weaker than that of Krypton, so Superman was able to lift heavy objects due to the difference in gravitational fields.

In *Superman* #58, Superman's powers are explained as follows:

> Everyone knows that Superman is a being from another Planet, unburdened by the vastly weaker gravity of Earth. But not everyone understands how gravity affects strength! If you were on a world smaller than ours, you could jump over high buildings, lift enormous weights . . . and thus duplicate some of the feats of the Man of Steel!

Which leads to our second basic question about Superman: How strong must Krypton's gravity have been to endow Superman with such incredible strength? Answering this question requires we first answer another: How massive was the planet Krypton that it had such high gravity?

Superman appears to weigh approximately 100 kg (220 lbs). An athlete in top physical condition can lift his own body weight. Running and throwing a heavy object might not be so easy. For our study, we're going to assume that Superman is 1,000 times stronger than an ordinary Earthman. That would mean he could lift 100,000 kg or approximately 220,000 pounds. This is approximately the weight of three filled semi-trailer trucks or a DC-9 airplane without fuel or passengers. Cranes used to construct bridges can handle about that weight, so we'd have a Superman still well within the bounds of human imagination. Such strength would even enable him to leap a mile with one jump, thus approximating flying in the eyes of most people.

The force necessary to lift an object on a planet is equal to the mass of the object multiplied by the gravitational force present on that planet. Thus, a human who could lift 100 kg on Earth could lift 600 kg on the Moon, which has one-sixth the gravity of Earth. Which would imply that for Superman to be 1,000 times stronger on Earth than he is on Krypton, Krypton would have to be 1,000 times as massive as the Earth.

Earth's gravity is 9.8 meters/sec$^2$, or for simplicity's sake, 10 meters/sec$^2$. Multiplying that number by 1,000 gives us the gravity of Krypton, 10,000 meters/sec$^2$.

Could a planet exist with such a gravitational field? According to Brother Guy Consolmagno of the University of Arizona, a planet with even fifty times the gravity of Earth "is essentially impossible to construct, given the physics of solid matter as we understand it."[8]

Put in even simpler terms, "a body with . . . a surface gravity of 10,000 m/sec$^2$ would have a mass of $6 \times 10^{33}$kg . . . , which would be 3,000 times the mass of the sun."[9]

According to the basic laws of physics, Krypton is impossible.

Moreover, for people resembling us to live on Krypton, they'd

[8]In an email to the authors, 5/06/01.

[9]Robert Rostrom, in an email to the authors, 5/06/01.

need muscle and bones 1,000 times stronger than human muscle and bone. No such material exists to create bone or muscle, or the complex internal organs necessary for life as we know it.

On a planet with gravity 1,000 times that of Earth, would it be possible to send a rocket ship, especially a small one as seen in numerous issues of *Superman* and *Action Comics*, to Earth? The escape velocity (the speed necessary to break the gravitational pull of a planet) of Krypton would be enormous, approximately 11,000 km/second. That's about 1/30 the speed of light. No chemical reaction in the universe could produce enough energy necessary to achieve such velocity.

In the 1960s, the explanation for Superman's powers was revised: his super strength, ability to fly, and more came not only from the high gravity of Krypton but also from growing up under a yellow sun instead of a red one. Unfortunately for Superman, light is light. The light from a red sun would merely have a smaller occurrence of high frequencies than the light from a yellow sun. Infrared light would be more common, but that's about it. Red star or yellow star, Superman's powers would be the same.

Superman is one of the most fascinating characters in comic books, and he's one of the most recognizable characters on Earth. He's one of those people we wish could exist, but doesn't. As we've shown, visitors from other planets are possible. Superman's not.

Chapter 2

# Rays—Cosmic and Gamma

## The Fantastic Four and The Incredible Hulk

### Humble Beginnings

**Superhero comics experienced** a sharp decline in sales during and after World War II. Much of their audience was drafted, and when the boys came home from the war, the concept of a flying man who couldn't be harmed by bullets or a caped crusader who battled crooks without using a gun seemed silly. Comic sales faltered as publishers looked for new gimmicks to sell their magazines.

Funny animals, a mainstay of the Disney comic book chain, appealed to younger children and continued to sell well. Thus, other publishers, including DC, tried publishing comics in the same vein, featuring characters such as Mighty Mouse, Peter Porkchop, and the Fox and the Crow. War comics were another category that gained a loyal following, as did western comics, but the two areas that experienced the greatest leap in sales were crime and horror comics.

Graphic violence in comics raised the specter of government intervention in the comic book field. After Senate hearings in the 1950s, the few remaining comic book companies created a trade association called the Comics Magazine Association of America. The Comic Code of America seal on a comic book insured that the issue featured safe reading matter for children. Comics without the CCA code didn't stand a chance. Union truck drivers wouldn't handle them, stores wouldn't sell them. So the middle years of the 1950s became a time of bland superheroes and funny animals.

Fortunately, DC Comics had stayed away from gruesome horror or crime comics. Their line was aimed at mainstream middle America. Titles in the 1950s included *Star Spangled War Stories*, *Our Army at War*, *All Star Western*, *A Date with Judy*, *Congo Bill*, and *The Adventures of Bob Hope*. The company's mainstay superhero comics, *Batman* and *Superman*, and related titles like *Action Comics* and *Detective Comics* were free of excessive violence and had little problem satisfying the CCA.

By 1956, DC was the king of the superhero field. They offered two big heroes in Superman and Batman, as well as a strong supporting cast of friends and sidekicks, including Jimmy Olson, Lois Lane, Robin, Aquaman, and Wonder Woman. That domination increased with the appearance of two new titles, *Showcase* and *The Brave and the Bold*. Each comic served as a launching pad for new heroes or characters revived from the Golden Age of comics, complete with new origins and costumes.

The first such revival was the Flash, who appeared in *Showcase* #4, September 1956. After returning in *Showcase* for several more adventures, the Flash earned his own comic in 1959. In the years that followed, he was joined by the Green Lantern, the Atom, Hawkman, and the Challengers of the Unknown.

One of DC's most successful new titles was *The Justice League of America*, a book that featured a group of heroes including Superman, Batman, and Wonder Woman battling terrible foes in a story that filled the entire comic. According to a comic industry legend, Jack Liebowitz, one of the owners of DC Comics, was playing golf one day with Martin Goodman, the owner of Atlas Comics, and bragged about the incredible sales of *The Justice League of America*. Story has it that Goodman went back to his office after the game and told his editor, Stan Lee, to write a comic book just like the Justice League, featuring a team of superheroes. And that's how Marvel Comics began.

Goodman, at one time a major player in the comic book field, was the publisher of a small chain of eight comics that featured stories about giant monsters from outer space. Distribution, oddly enough, was handled by DC. Goodman's main man was Stan Lee, who served as both editor of the line and writer on most of the titles. Art was provided by freelancers who worked from Lee's scripts.

Lee put together the new comic book with the help of veteran comic book artist Jack Kirby. The first issue of the new series, titled *The Fantastic Four*, appeared in November 1961. It was the beginning of what became known as "the Marvel Age" of comics.

## A Fantastic Foursome

The group's origin story, as described by Lee, had scientist Reed Richards, his fiancée, Sue Storm, her teenage brother, Johnny, and Reed's friend, test pilot Ben Grimm, flying an experimental rocket ship into the edge of space. Bombarded by cosmic rays, the ship crashed, but all four members of the crew survived. However, each of them was forever changed by the hard radiation.

Reed Richards found that he could stretch his body to incredible lengths. Sue Storm was gifted with the ability to turn herself invisible. Later, she also learned to control an invisible force shield. Her brother, Johnny, became the Human Torch, able to become a living flame. And Ben Grimm, the test pilot, changed into a rock-like monster with superhuman strength, dubbed the Thing.

Acting as both editor and writer for the new series, Lee took the comic in directions never before emphasized in the field. For one, his characters had human traits and personalities. Johnny was a headstrong teenager who didn't like taking orders from his future brother-in-law. When not complaining about his grotesque appearance, Ben Grimm made cynical remarks about the world and his teammates. Sue Storm was young and romantic, while Reed Richards was a geek and stuffed shirt.

Another of Lee's changes was his use of continuing storylines in the comic. For the first time ever in superhero comics, the plot continued from month to month. Characters changed and matured.

*Continuity mattered.* While individual stories remained accessible, they all formed part of a larger whole, instead of being entirely separate. Month after month, Reed searched for a cure for Ben Grimm's condition. Ben met a blind girl and fell in love. Sue and Reed's romance had its share of bruises and crisis moments. Villains fought the team and were defeated, then returned six months later with new powers and new dreams of conquest.

It was all very different for comics. *The Fantastic Four* was more than a superhero comic book. It was a continuing soap opera, the "As the World Turns" of comics—and readers loved it.

What Lee basically did was change the dynamics of comic book relationships. In the past, there was little real interaction between Superman and Lois Lane, or Bruce Wayne and Vicki Vale. The status quo in most comics never changed. Nor was there any indication that it ever would. In the new Marvel titles, relationships were volatile and dynamic. Reed Richards and Sue Storm were in love, and despite numerous disasters and near-death experiences, their love lasted and continued until Sue and Reed married, in one of the major weddings in comic book history. Later, Sue even gave birth to a son, who needless to say was soon involved in the complex Marvel soap opera plotlines. While major changes weren't extremely common in Marvel titles, they did happen, often to the surprise of the readers.

Continuing stories swept through comics and destroyed the status quo. In time, all major comic companies found themselves writing stories with continuing plotlines and changing characters. It was the move that took comics back from children and returned them to teenagers and adults.

## Frankenstein's Monster—Marvel Style

The first issue of *The Fantastic Four* sold extremely well, surprising everyone at Goodman's publishing company, including Stan Lee. More important, the second issue posted strong numbers, as did the third and the fourth and every issue succeeding them. Lee suddenly found himself writing and editing the most exciting new comic in the field. A new logo was adopted for the comic: "The World's Greatest Comic Magazine!"

Seeking to capitalize on the success of *The Fantastic Four*, Lee came up with another concept, again done in collaboration with Jack Kirby. Lee called his new character the Incredible Hulk, a cheerful reminder of the names of stories Marvel once published.

The new comic appeared in May 1962, approximately six months after the debut of *The Fantastic Four*. The first issue of *The Incredible Hulk* combined elements from Mary Shelley's *Frankenstein* with the story of Ben Grimm, the Thing, from *The Fantastic Four*. Dr. Bruce Banner, the inventor of the "gamma bomb," was caught in the bomb's first test blast due to the machinations of a communist spy. Banner seemed unharmed by the gamma rays until he grew agitated or his temper rose. Anger turned the mild-mannered Banner, 5'9'' and 128 pounds, into a gigantic man-monster with superhuman strength and a mean attitude. The Hulk stands at 6'6'' and 7' tall (depending on his color) and weighs between 900 and 1,040 pounds. He's capable of lifting between 90 and 100 tons. And he was grey, then green (a printer's problem that later became a plot line, with the color variations attributed to different Hulk personas).

Atomic bombs had served as major threats in comic books since the end of World War II. Photos of the terrible destruction caused by the bombs at Hiroshima and Nagasaki were in numerous textbooks, and stories about atomic testing were commonplace. Facts concerning the devastation and destruction caused by the bombs were everywhere. Whereas a TNT explosion had temperatures of a few thousand degrees, the temperature in an atom bomb mushroom cloud was measured in millions of degrees.

No one, including many scientists, was sure of the effects of hard radiation on animal tissue, and dozens of B-movies and comics featured radioactive monsters. Nor were people sure of the effects of radiation on the human brain.

The Hulk took paranoia about the bomb to the next level. The Hulk was the terrible by-product of atomic bomb radiation. Making things worse, he was both a good man and a horrifying monster sharing the same body. The concept of dual personalities in the same body was already a cliché, but it served Lee and Kirby well. The Incredible Hulk was blood brother to the Amazing Colossal Man and nephew to Dr. Jekyll and Mr. Hyde. The Hulk was a rampaging monster and an innocent victim combined in one.

The one important difference with Lee and Kirby's creation was that Bruce Banner's alter ego, aka the Incredible Hulk, was the hero and not the villain of the comic book series. Banner hated turning into the Hulk, but there was little he could do about it. Since the Hulk was technically a hero, albeit a confused and often misguided one, he constantly found himself in situations where he used his amazing strength to battle evil menaces like the Toad Men, the Metal Master, and the Leader.

Oddly enough, though *The Incredible Hulk* sold quite well and was another hit for Marvel, his comic was canceled after six bimonthly issues. Martin Goodman's comics were distributed by DC comics, and their contract only allowed for Goodman to publish eight comics per month. With Stan Lee coming up with idea after idea, Goodman just didn't have enough comics to accommodate all of his new heroes. *The Incredible Hulk* was canceled so that another new character could appear in his own book: *The Amazing Spider-Man*.

But the Hulk was too popular and too good a character to just abandon. He soon appeared as a guest star in *The Fantastic Four*, battling the Thing. Stan Lee was weaving together his own universe. The characters in that universe, both heroes and villains (and supporting characters), were used in all of Marvel's comics. So it wasn't unusual for a villain battled by the Fantastic Four to turn up later fighting Ant Man or Spider-Man. When Spider-Man went looking for a job, it seemed quite natural he'd apply for work with the Fantastic Four. It was just another example of how Lee was creating not merely a bunch of heroes but an entire elaborate world in which they played. The notion had been used before in DC Comics, with Batman and Superman sometimes teaming up to fight common foes, but it was more a novelty than a continuing theme. DC's characters stayed within their own domains. Marvel's characters all lived and fought in the same city (New York) and ran into each other quite frequently. It was a gimmick, but it worked.

In an effort to get around the DC distribution problem, Lee turned comics already being published, like *Journey into Mystery*, *Tales to Astonish*, and *Tales of Suspense*, into superhero vehicles for new characters like the Mighty Thor, Ant Man, and Iron Man. Instead of monster stories, the Marvel line concentrated entirely on superheroes.

The Hulk soon reappeared as a cofeature in *Tales to Astonish*. Once Marvel ended its distribution deal with DC and signed with Curtis Circulation in the late 1960s, the Hulk once again got his own comic, starting in April 1968. He's been published steadily ever since, remaining one of Marvel's most popular characters. The Hulk has also appeared in cartoons and on his own live-action TV series, and he's scheduled for a major film, directed by Ang Lee, to be released in 2003. He stands as another major success for the team of Stan Lee and Jack Kirby.

But is he the least bit believable in scientific terms? What exactly is a "gamma bomb," and could its radiation, or any type of radiation, affect a man's physique? Can radiation make someone smarter? What about the Fantastic Four? Can cosmic rays turn ordinary people into superheroes? Or make them invisible?

And, leaving aside paint, dyes, and envy, is there any way to turn someone green?

## The Perils of Technobabble

Any book dealing with science in entertainment needs a place to discuss technobabble. Invented in 1981 to describe the technical jargon used on *Star Trek*, the word refers to any outlandish technology, whether futuristic, alien, or of otherwise undetermined origin.

Technobabble is best when it uses familiar words in new and unfamiliar ways. Like most double-talk, it should sound believable and perhaps even bear some small relationship to the technology it supposedly describes. But it needs to remain separate from anything to do with the true meaning of the terms employed.

The word technobabble has become so popular that it's been accepted in mainstream vocabulary as meaning any definition (particularly in the computer field) involving jargon that is meaningless to someone outside the particular sphere of endeavor. Thus computer technicians are accused by politicians of speaking in technobabble. In its most general use, technobabble stands for "scientific gibberish." It's not a word often used as a compliment.

Comic books, particularly superhero comics, have always been a friendly technobabble environment. Most scientific explanations of super powers depend on a certain amount of it. DC Comics, however, were much more science-oriented than most of their competitors,[10] in part because DC editors Mort Weisinger and Julius Schwartz were longtime science fiction fans, and they used a number of science fiction writers, including Edmond Hamilton, Otto Binder, and Manly Wade Wellman, to write their scripts. These authors, schooled in writing for critical SF fans, knew how to stretch scientific points so that they almost seemed to make sense. At other publishers, technobabble ruled. Scientific explanations were fine if they fit the concept of the story. But the story was the thing.

And in this lies the difference between the Incredible Hulk and the Fantastic Four.

By the second issue of his comic, the Hulk was green. Complicating the story was Bruce Banner's relationship with Betty Ross, daughter of army general "Thunderbolt" Ross, sworn to destroy the Hulk. Another problem faced by Banner was that he was every bit

[10]"I always included lots of short science features in my comics," said Julius Schwartz in a recent conversation with the authors (8/14/01). "A number of people have come up to me over the years and said my comics helped them pass high-school science courses!"

the quiet intellectual often found in comics who suddenly must deal with the dark, primordial side of his psyche.

Decades of rewriting and reexamining the Hulk's origin have never been able to wipe away one word from the Green-Skinned Goliath's origin story. Details (in horrifying numbers) have been added and rewritten and expanded until we know everything and more about Bruce Banner's life before the fateful experiment. Still, Bruce Banner changes into the Hulk due to his exposure to *gamma* radiation. The secret of technobabble is never to use a word that appears regularly in science, especially in meaningful relationship to the subject you're describing. That's the problem with gamma rays. They're real, but they don't turn people into monsters.

Gamma rays are, like light, a form of electromagnetic radiation. All electromagnetic radiation travels in waves and is classified by its wavelength, the distance between the waves. There are many types of electromagnetic waves in the universe, only a very small amount of which are visible to the human eye. Gamma rays are invisible to us, as are X-rays, radio waves, ultraviolet light, and infrared radiation. Gamma rays are extremely high energy waves and thus have very short wavelengths on the electromagnetic spectrum. They are about one millionth the size of light rays, which are one thousandth the width of a human hair.

Gamma rays are emitted by radioactive substances like Cesium-137, Cobalt-60, and Uranium-235. Along with alpha and beta rays, gamma rays are emitted when a nuclear explosion takes place. Because Bruce Banner is described as a *nuclear* scientist in the first issue of *The Incredible Hulk*, there's no question the gamma ray bomb (or G-Bomb) being tested is a high-level nuclear device. It's also clear that after Banner's exposure to the full force of the gamma rays, everyone at the test site is astonished that he's still alive. Especially since he's so radioactive he starts a Geiger counter chattering.

And for good reason. The results of high exposure to gamma radiation are well known to scientists and have been known since the dawn of the atomic age. The unit of measurement used to measure radiation dosage is the rem, which stands for roentgen equivalent in man. These units represent the amount of radiation that will produce a certain amount of damage to human tissue. Gamma radiation doesn't cause transformations in human beings. Instead, it causes burns, cancer, and death.

If anything, a G-bomb or gamma ray bomb would only be given such a name if it produced more hard radiation—gamma radiation—than a normal atomic bomb. Even without pushing our numbers, if Bruce Banner received the full brunt of the gamma rays emitted from his bomb, if his body set a Geiger counter crackling hysterically, he'd have received a dose of 1,000 rems or more. Any reading of 800 rems or over is fatal, with the victim dying painfully within several days of the exposure.

There is no known cure for radiation sickness. Signs of the high radiation doses are nausea, vomiting, headache, loss of white blood cells, hair loss, damage to nerve cells, and damage to the cells lining the digestive tract. Radiation sickness also reduces the production of platelets in the blood that aid clotting, so people with high rem experience blood hemorrhaging.

If Banner was exposed to somewhat less than 800 rems but more than 400 rems, his chance of living more than a few days is somewhat increased to about 50 percent. However, he'd be at high risk for the rest of his life for leukemia (blood cancer), thyroid cancer, lung cancer, and numerous other cancers. Many radiation victims also experienced premature aging.

What's true for the Hulk is equally true for the Fantastic Four. Instead of being exposed to "gamma rays," Reed, Sue, Ben, and Johnny are exposed to *cosmic* rays. That's hard radiation that comes from space. We're not exactly sure from where (supernovas, also a source of gamma rays, are a popular choice these days, but no one is definite). But the source is less important than the effect.

Hard radiation doesn't transform living cells. Nor does it cause certain glands to overreact and mutate human bodies. When charged particles travel through human tissue, they strip electrons from atoms and molecules, thus destroying their ability to function. Hard radiation in large doses only does one thing, and it does that well: it kills. If Dr. Bruce Banner was exposed to high levels of gamma radiation at a nuclear test range, he would have died within two days of the event. No matter whether he's green or gray, smart or stupid, The Hulk just doesn't make sense according to his origin story. Neither do the Fantastic Four. Marvel's earliest heroes just aren't possible and they never were.

We can't fix the Fantastic Four. Their powers are too bizarre to explain, no matter what their origin. However, if we go back and rewrite the Hulk's beginnings—an accepted principle in comic books known as retconning—it's easy to come up with a much more scientific, almost believable version of the Giant Green Goliath. We like to think of him as the GFP Hulk.

## The GFP Hulk

Dr. Bruce Banner was a typical geeky nuclear scientist of the early 1960s. He's the guy who always had sand kicked in his face in those Charles Atlas ads. Though Bruce was the valedictorian of his high school and college graduating classes, he never could get a date because he was a wimp. It's the same old story of the smart guy with glasses losing out to the no-brains but all-brawn football hero. Bruce was a man born out of time. If he'd only waited to grow up in the 1990s, he could have had it all as a super hacker or web designer. But Bruce couldn't predict the future and the eventual triumph of the slide-rule set, so he took action—dangerous, life-threatening action that changed him from a mere 128-pound weakling to a massive green monster known as the GFP Hulk.

Instead of majoring in nuclear engineering and gamma ray bombs in college, our Bruce Banner studied chemistry and biology. Bruce knew that high-level radiation destroyed cells and didn't mutate anything. He had no desire to shorten his lifespan to a few days. So Bruce studied the inside of the human body instead of the inside of atoms.

In the course of his work, Bruce examined the adrenal glands, a pair of glands located above the kidneys that secrete hormones directly into the bloodstream. These glands are only an inch or two long and weigh a fraction of an ounce, yet they produce over three dozen different hormones. The adrenal glands are controlled by the pituitary glands and have important effects on a person's physical development and growth.

The outer section of the adrenal glands secretes hormones that control body shape, hair growth, and the way food is used in the body. These hormones are known as steroids. The inner sections of the adrenal glands produce chemicals that react to fear and anger.

Investigating further, Bruce learned that steroids could be produced artificially. Anabolic steroids are artificial versions of testosterone, one of the most important steroids. Along with helping develop male sexual traits, anabolic steroids cause muscles to grow. These artificial steroids were a favorite among body builders for developing bigger muscles until they were removed from the marketplace in the 1980s and made available only by prescription.

In true comic book hero fashion, Bruce Banner devoted himself to developing a super version of anabolic steroids. When taken, these pills caused huge muscle growth in his body, easily pumping up Bruce from a 128-pound weakling to a 940-pound, seven-foot-tall behemoth described by women on campus as "an incredible Hulk!"

Bruce, as is too often the case of scientists in comics, acted a little too hastily in testing his new pills. He didn't pay enough attention to the side effects of steroids. Among the many dangers of using steroids was an irreversible loss of scalp hair. Thus the Hulk was completely bald. Worse, anabolic steroids interfered with the normal production of testosterone in the human body, which lowered Bruce's sex drive and reproductive drive. While Bruce might have been idolized by many of the girls on campus, he felt more impotent than incredible.

Yet, loss of his sex drive wasn't the worst of Bruce's problems. Anabolic steroids affected the limbic system. That's the part of the brain that influences moods and is involved in learning and memory. Steroids led to major mood swings in Bruce, including feelings of rage and depression. Aggressive behavior was another side effect of the steroids. The Hulk soon became known as much for his titanic rages as for his titanic muscles.

It was perhaps in one of those mindless rages that Bruce performed one more drastic experiment on his ravaged body—an act that would forever mark him as "this man, this monster" and would cause him to be dubbed "the Green Goliath" of the comic book world.

Bruce went to see a lecture by Eduardo Kac, an assistant professor of art and technology at the School of the Art Institute of Technology. Kac discussed his controversial transgenic art and, in particular, his most controversial creation, Alba, the GFP (green fluorescent protein) bunny.

Alba is *real*, and was created by French genetic researchers through zygote microinjection. They removed fluorescent protein from a species of jellyfish, modified the gene so that it glowed more brightly, and then inserted the gene (called EGFG, the enhanced green fluorescent gene) into a fertilized rabbit egg cell that grew into Alba. The green gene was present in every cell of Alba's body. When Alba rested beneath a black light, the rabbit glowed green.

The green gene did have important uses. It was used to code specific genes or proteins. When the protein was active, the fluorescence could be detected under a black light. In reality, scientists have used this tracing ability to observe anti-cancer genes by black light. In the future, doctors hope to use the green gene method to help locate cancer cells in humans.

In the comic, Banner was so impressed with Kac's work that he decided to see if he could duplicate the GFP process with a human being. With his judgment clouded by high steroid use, Bruce decided to use himself as the first test subject. Needless to say, the GFP gene reacted with the synthetic steroids in Banner's system with dire results. The results were predictable. Whenever Banner got angry or went into a rage, the artificial steroids changed him from an ordinary man into the gigantic, dull-witted Hulk, and the steroids activated the GFP gene in Banner's body, causing his skin to glow green.

Since his graduation from college, Bruce Banner, one of the great geniuses of the twentieth century, devoted his life to fighting drug and substance abuse. In his human persona, he was at the forefront of developing new medications that would hopefully help people realize the adverse consequences of their actions. In his transformed state as the GFP Hulk, he hunted and smashed drug lords, crime syndicates, and people who were cruel to animals.

Chapter 3

# The Dark Knight

## Batman

### A NonSuper Superhero

**One of the** true icons of comic book culture is Batman, a superhero without super powers. The scourge of the underworld, Batman is a spectacular crime fighter with a dazzling array of weapons and gadgets. Unlike most comic book heroes who are gifted with extraordinary powers, Batman is an ordinary man who develops his skills through training and hard work. A master detective, Batman is one of the few superheroes who outthinks as well as outfights his opponents.

The creation of artist Bob Kane, Batman first appeared in *Detective Comics* #27, May 1939. Like Superman from a year before, the costumed crime fighter caught on quickly with the reading public, and within a few years was starring in his own comic as well as continuing to appear in *Detective Comics*.

Unlike Superman, however, Batman wasn't unique in comics. Before he debuted, a number of noncostumed heroes appeared in the pages of *Detective* and *Action Comics*. Soon after Batman's appearance, a number of very similar costumed heroes with secret identities joined the ranks of comic book characters, yet none ever achieved the same level of success as Batman. None ever became an American legend recognized throughout the world. What made Batman so special?

For one, his look was unique. Batman's name came from his appearance. He looked like a bat in human form. With his cloak and hood, he looked like no other character in comics. With his bright red and blue outfit, Superman was an all-American hero. Clad in dark colors and wearing a mask while fighting crooks, Batman was a creature of the night. To use a catchphrase invented many years later, the early Batman truly was a "Dark Knight."

Then there were Batman's roots. Unlike Superman, who was defined by his parents' noble sacrifice, Batman was the product of murder. Batman's tragic history gave him a depth of character unequalled by most superheroes. As pointed out by comic book historian Les Daniels, Bob Kane created Batman months before ever considering the character's origin.[11] Kane was more concerned with his hero's look than with his history. The story of how young Bruce Wayne's parents were killed before his eyes by a petty criminal, inspiring the boy to devote his life to fighting crime, didn't appear until December 1939. Kane invented Batman, but it was Kane and the writer Bill Finger who together devised Batman's background. The right look and the right history combined to make Batman a compelling character.

Equally important in shaping Batman was the decision in late 1939 by newly appointed DC editorial director Whit Ellsworth to keep the actual violence in Batman stories to a minimum. Early adventures in *Detective Stories* had Batman using a gun to dispatch villains. Ellsworth wanted DC comics to be kid-friendly and reasoned that too much violence would alienate readers. Within a year, guns were gone and Batman was capturing, not killing, criminals.

The next major step in Batman's evolution came with the addition of a kid sidekick, Robin, in *Detective Comics* #38, April 1940. Bill Finger, who was writing the scripts for the series, complained that Batman had no one to talk to. Bob Kane obligingly created Robin.[12] The Boy Wonder added dialogue to the comics and also gave read-

[11]Daniels, Les. *DC Comics, Sixty Years of the World's Favorite Comic Book Heroes*, Bulfinch Press, New York, 1995, p. 32.
[12]Ibid., p. 36.

ers a character their own age. Robin proved to be a wise move. The circulation of *Detective Comics* nearly doubled after the addition of the teen hero, leading to a proliferation of teenage superhero assistants over the next two decades.

The final key to Batman's success was the bizarre crew of villains he battled each month. Finding worthy opponents for Superman required enemies with incredible powers or super science. Batman, a crime fighter who used his intelligence to battle crime, merely needed criminals with a good gimmick to make them worthy opponents. Bob Kane, Bill Finger, and writer/artist Jerry Robinson created a wild rogues' gallery for Batman that was spectacular even by comic book standards. Top-notch villains included the Joker, Catwoman, Two-Face, the Penguin, the Riddler, and many more.

Appearance, history, a teen sidekick, and intriguing villains make Batman one of the most popular superheroes in history. During Batman's more than sixty years of comic book stardom, the formula has changed on occasion. For example, the original Robin grew up and needless to say became a crime fighter. A second Robin died. However, his replacement fights by Batman's side today.

Other writers and artists following the team of Bob Kane and Bill Finger reshaped Batman to fit the times. Most notably, Frank Miller turned Batman into a darker, grimmer, more realistic character in his retelling of Batman's origin in the 1980s with "The Dark Knight Returns." Miller's stark imagery served as a major influence for Tim Burton's film *Batman*. Today, Batman continues to shine as one of DC Comics's greatest stars, with his adventures highlighted in a half-dozen comic books every month.

A non-super superhero, fighting non-super criminals. Where's the science? Just keep reading.

## The Science of Batman

Unlike Superman, Batman wasn't born with super powers, nor did a friendly alien like the Green Lantern give Batman super powers. An explosion didn't douse him with chemicals as in the case of the Flash. Batman didn't fly a homemade rocket to outer space like the Fantastic Four, nor did he witness a gamma ray explosion up close like the Incredible Hulk.

Batman is a self-made hero. As explained in numerous stories, including "How to Be the Batman," *Detective Comics* #190, Batman spent years training in a gym to become a perfect acrobat. He directed his entire education toward scientific crime fighting. Bruce Wayne trained his mind in much the same manner in which he trained his body. As he stated in *Detective Comics* #190, "I've got to know science thoroughly to become a scientific detective." There's little question that Wayne succeeded beyond his wildest dreams.

Consider "The Amazing Inventions of Batman," as discussed in *Batman* #109, August 1957. In this story, Batman and Robin use portable jet packs to fly between buildings; use a heat ray to detonate dangerous boxes of explosives floating in Gotham City harbor; and use a flying camera to spy on crooks planning a robbery. Was Batman using real-life science or merely 1950s pseudoscientific nonsense?

The first accurate prediction of a portable flying pack was made in 1928 in the novel *The Skylark of Space* by E. E. Smith, also serialized in *Amazing Stories*, August through October 1928. The first issue of the magazine featured a cover with a man flying while wearing a rocket backpack. The same issue of *Amazing Stories* also featured the first Buck Rogers story. Although the cover had nothing to do with Buck Rogers, the flying backpack illustration and Buck Rogers were forever linked by inaccurate research as that "crazy Buck Rogers stuff."

The writer of "The Many Inventions of Batman" was Edmond Hamilton, a friend of DC editor Julius Schwartz and a longtime comic book scriptwriter. Hamilton had been writing science fiction stories since 1926, and there's little question that he read the August 1928 issue of *Amazing Stories*. Most likely, it served in part as Hamilton's inspiration. But, quite possibly, so did real science.

In the 1950s and 1960s, magazines like *Popular Science* and *Popular Mechanics* ran several articles about portable jetpacks being developed by scientists trying to come up with new methods of transportation. The man most often mentioned regarding such devices was Wendell F. Moore, a scientist who worked for Bell Aerosystems during those years. Moore dealt with small rockets fueled by hydrogen peroxide. According to several news accounts, he came up with the idea of a man flying by the use of small rockets on his back one evening while doodling.

Moore's doodles turned real in 1960 when the Army Transportation Command awarded Bell Aerosystems a contract for $150,000 to develop a Small Rocket Lifting Device. The Army wanted a practical machine for improving troop mobility. Moore built his rocket belt and on April 20, 1961, an associate of his at Bell Aerosystems, Harold Graham, flew 112 feet outdoors using the rocket belt.

Unfortunately, the Bell jetpack was highly impractical. The invention was little more than a high-powered rocket strapped to a man's back. The jetpack used pressure from liquid nitrogen to force hydrogen peroxide into a catalyst chamber where it reacted with silver screens coated with samarium nitrate. The mix created a jet of very hot, very high-pressured steam that provided the thrust that lifted the user into the air. One wrong move and the pilot was badly burned by the steam. Equally dangerous, the flier had to use his own legs as landing gear. In addition, the jetpack made an incredibly loud noise when in operation.

Despite all of its flaws, the Bell jetpack fascinated the public. The device was demonstrated numerous times around the world. It was shown on television shows and in air shows, and was even featured in the James Bond film *Thunderball*.

The Army never used the Bell jetpack for the simple reason that it could only carry enough fuel for a twenty-second flight. When Moore died in 1969, the jetpack was retired from use. However, the idea of a personal flying device refused to die. An August 2000 news release[13] described the Solo Trek XFV, made by Millennium Jet

---

[13]"Personal Jetpacks Take Off," www.howstuffworks.com/news-item149, August 10, 2000.

Inc., a vertical one-man jet that could fly up to 80 miles per hour and for 150 miles before refueling. Chalk one up for Batman.

Flying police weren't anything unusual in Gotham (a.k.a. New York City). The first police helicopter patrols in the world began in Manhattan in 1948. Batman merely took a proven idea and pushed it one step farther. In "The Many Inventions of Batman," the Dark Knight used a flying camera to spy on a criminal scientist and his gang. While aerial surveillance was in its most primitive stages in the 1950s, it was an idea that was evolving. In the comic book adventure, the criminals steal Batman's invention and use the flying camera to locate an armored car traveling on the highway. Any resemblance to a certain car chase involving an ex-football player was purely coincidental. Just remember, Batman predicted it first!

What about the heat ray used to detonate explosives in the water? Lasers can be traced back to Albert Einstein's 1917 theories. The first microwave laser was built in 1954, three years before the Batman story took place. The first optical laser was invented three years after the story was published. Batman was merely taking existing science and projecting it forward a few years.

In our time, small, high-powered diode lasers are often used in delicate surgical procedures, but could be wielded as a weapon if necessary. Still, whatever damage they could cause, lasers aren't nearly as effective as low-tech weapons like guns or knives. The armed forces have conducted tests with much more powerful lasers, but the results of these tests aren't available to the general public. The most common use of a laser in warfare is as a powerful light gun, causing major eye damage to distant enemy forces. The use of lasers to blind people in warfare has already been banned in an international treaty.

The most powerful tool used by Batman in his war against crime was his utility belt. On it, he kept a number of tools and devices to help him battle criminals and solve mysteries. The multifaceted utility belt has become a part of American pop culture. Some computer hackers hang all sorts of electrical equipment like pagers, personal organizers, pocketknives, flashlights, tool kits, and even miniature computers from their belts. Needless to say, the hacker nickname for such a belt is a "bat belt."

According to the first *Giant Batman Annual* published in 1961, the following items are contained in Batman's utility belt:

Explosives
Infrared flashlight
Smoke capsule
Fingerprint equipment
Miniature camera
Pass keys
Tiny oxyacetylene torch
Gas capsule

Batman's silken rope is described as being drawn out of the belt lining as a fisherman's line is drawn from a reel.

It was a fascinating list for the time. Miniaturized items weren't readily available in a world before microcircuits and computer chips, but still, were the items outrageous or merely projections of what science promised for the future?

Do we even need to mention miniature cameras? Every appliance, electronics, and camera store in the United States has a full stock of miniature cameras.

There's also the Fraser-Volpe Co. M.I.C.E.—miniature integrated camera eye—which is a wireless high-resolution camera system. It can be used as a standalone camera, and it can also be attached to all sorts of optical devices such as binoculars and rifle scopes. The system lets the optical device function normally while it transmits realtime videos back to a command center. It's a device many people thought only appeared in *Mission Impossible*, but it's real. There's no question that it would be part of Batman's arsenal.

Next, we have pass keys. These keys are part of any respectable burglar's equipment and something that every crime fighter needs in his war against the underworld. Pass keys or picklocks are legal in most states, but it *is* a crime to be caught carrying such tools if there is clear indication of criminal intent. Most professional thieves know better than to carry picklocks with them, since almost any thin piece of metal (or sometimes plastic) is all that is necessary to open most locks.

What's true for criminals we must assume is true for Batman as well. In his utility belt, he probably carries a small set of "jiggler" keys, very thin keys that can be inserted into most locks to jiggle the tumblers, and a few master keys, general all-purpose keys that slide easily into many locks. Along with keys, Batman carries several lock picks and tension wrenches. These are easily made from pieces of spring steel, including piano wire and hacksaw blades. Using these few tools, the Dark Knight can enter nearly any apartment or building with ease.

Electronic locks are more of a challenge, but a little ingenuity and an electronic coding device will work wonders. Automobile security isn't any more challenging for crime-fighters than it is for criminals; most door locks can be opened with a thin piece of wire. Despite advertisements to the contrary, the famous "club" can be picked by most car-jackers in seconds. Besides which, most steering wheels are vulnerable to attack with or without an attached club.

Do you think that a miniature oxyacetylene torch is impossible? Most units can't fit on a desk, much less on a man's belt. Still, nothing is impossible in modern times. An MEC Midget Torch weighs just six ounces and isn't much longer than a man's hand, and the tip is narrower than a human finger. Maybe it's not a comfortable fit in the front of the utility belt, but it's portable, so Batman could easily attach it to the side or back.

When Batman faces desperate odds, a gas capsule makes the fight a lot fairer. Modern policemen often have the same problems when trying to disperse an unruly crowd. Following the Dark Knight's lead, they also can use gas, and not some unwieldy canister that needs to be fired from a rifle. A 4-inch by 1-inch Punch II police-strength pepper spray[14] will stop even the most violent criminal. A one-second spray aimed at the eyes causes temporary blindness. It also induces choking, coughing, and nausea. Violent drug

[14]As advertised on the internet.

addicts and psychotics aren't bothered by Mace or tear-gas products, but they're not immune to oleoresin capsicum, the active ingredient in pepper spray. Batman's most violent enemies, human monsters like Bane, might feel no pain, but they're not immune to this less-than-lethal but highly effective aerosol weapon.

Smoke grenades are easy. They're available in the mini-size of 1¼ × 3–inch length (fits easily into a small pocket or container), and they generate 22,000 cubic feet of smoke. This is enough smoke to cause a lot of confusion during a fight. A large smoke grenade is 1½ × 6 inches and will generate a white-gray cloud measuring some 40,000 cubic feet. Both seem adequate for Batman's needs.

Infrared lights are 3.8-inch-long cylinders that produce a 5-foot-wide circle at 25 feet. They can provide infrared illumination for up to a hundred yards. The light lasts up to eight hours using one lithium battery. The entire light weighs 3.5 ounces.

We still need a dependable fingerprint kit. Most small kits aren't very useful, and partial prints taken at the scene of a crime usually prove to be unidentifiable when examined in a lab. The way to avoid this is to have a top-notch modern fingerprint outfit at the scene of the crime.

The Latent Print Developer Kit from Criminal Research Products Inc. offers a full line of fingerprinting products that will work on the scene as well as in a lab. Batman could carry tiny vials of the many types of latent print powder to identify his enemies. Antistatic latent print powder neutralizes static electricity from plastic surfaces. Atomic Brand latent print powder adheres to the actual fingerprint secretions, thus reducing smearing and ridge destruction. Silver black latent power is used when regular powder doesn't provide sufficient contrast. Zinc Print latent powder is used for developing latent prints on greasy, zinc-plated items such as vending machines and change boxes. Magnetic latent print powders are used specifically on paper, cardboard, wood, glass, plastic, and other non-ferrous surfaces. Safe-Cracker latent print powder is used on all metal surfaces such as safes, file cabinets, vaults, etc. Special combinations of these powers make finding fingerprints on just about any surface a thousand times easier than it was in the 1960s.

The last resource in Batman's utility belt is probably the most dangerous. Explosives are not meant to be carried in a belt pocket for many hours at a time. Even TNT, the most popular explosive of the twentieth century, can be deadly due to changes in the weather.

Lawrence Livermore Laboratories in California runs a HEAF (High Explosive Applications Facility), where they experiment with new miniature explosives. At Livermore, scientists design new molecules with explosive properties. They extenstively test these explosive compounds using computer models before they synthesize them.

One discovery made at the labs is the compound LX-19, which has the highest explosive power of any compound discovered in the world. Unfortunately, the material is too unstable to use as an explosive. However, research compound LLM-105 is sixty percent more powerful than TNT and is much more insensitive to its physical surroundings. It seems reasonable to assume Batman carries several sticks of this compound in his utility belt.

Though not exactly a part of his utility belt, Batman's silken rope was another one of the Dark Knight's most powerful tools. Equipped with a grappling hook (easily bought for under $30 at most stores handling mountain climbing gear, as well as at many internet shops that specialize in ninja accessories), the slender line offered Batman a silent and secret method to scale buildings without being seen by his enemies inside.

Batman was usually shown swinging on a rope, not climbing one. Swinging was more dramatic, but, in real life, it was not very practical in a big city. Not to mention that a rope-swinging Batman and Robin made perfect targets for crooks armed with machine guns. It's much more likely that Batman used his rope primarily for climbing.

In the 1940s and 1950s, no one believed men could scale tall buildings using mountain-climbing gear. It was the stuff of Batman comics and nothing more. A crazy idea going nowhere.

Today, that attitude has gone through a complete reversal. One of the most popular illegal "extreme" sports is "buildering," the art of climbing a city building without using any mountain-climbing gear. Young men and women climb the walls of large structures in cities, using cement hand holes and building decorations as their only aids. Entire magazines are devoted to buildering, and climbers frequently post the best routes for climbing major urban skyscrapers on internet bulletin boards. It's just another example of life duplicating comic books—in this case, ordinary teenagers and young adults imitating Batman.

Nothing in Batman's utility belt is beyond the reach of modern science. Some of the items might have been futuristic in the 1960s, but all of them are available today, proving that the Dark Knight (and his writers) had a keen eye for future developments in crime-fighting techniques.

Similar examinations of the Batplane, the Batmobile, and even Batman's huge crime lab located in the cave beneath Wayne Manor yield the same result. In the late 1950s, Batman's crime notes were on file cards, with duplicates of the cards kept on microfilm. As times changed, so did Batman's filing method. A computer database came into use. As computers grew more complex, so did the database Batman used. By the late 1990s, his computer was tied to those of major law enforcement agencies throughout the world, providing Batman with up-to-the-minute information about criminals anywhere in the world. What was once the domain of comic books has become part of the real world of crime fighting. Except for his vigilante methods and costume, Batman could be one of today's lawmen.

## The Gotham City Earthquake

One of the biggest and most complex stories involving Batman was in 1998 and 1999 in a multi-issue, multi-character crossover called "No Man's Land." The beginning of the story, labeled "Cataclysm," had an earthquake strike Gotham City. The earthquake destroyed Bruce Wayne's manor, as well as the Batcave and Batmobile. City Hall and the main Gotham police station were smashed, and much of the city went up in flames. Afterwards, the city was cut off from the outside world, and anarchy reigned as Batman and a few others tried to save what remained of the once proud metropolis.

Gotham City bears an uncanny resemblance to New York City. It's suffered plenty over the years as one criminal mastermind after another has made the city a living hell. But nothing compares to being hit by a major earthquake measuring 7.5 on the Richter scale. Pure imagination, or grounded in science and reality?

Definitely reality. In this case, comic book reality predicted real life, as on January 17, 2001, Manhattan and Queens experienced a minor earthquake registering 2.4 on the Richter scale. The location of the epicenter was somewhere on the upper east side of New York City about four miles down.

Although it is the West Coast that is famous for its earthquakes, the East Coast isn't immune to such tremors. New York City is located in the middle of a tectonic plate. It's not on the edge of a plate, as is the case in California. The New York City fault is a continental rift, a break in the rock caused by the collisions of continents 400,000,000 years ago. This rift runs through the Bronx, down the East River, through Staten Island, and down to Charleston, South Carolina.

Rifts have a tendency to produce earthquakes. This puts New York City at risk. Plus, there are numerous fault lines in northern Manhattan running both north-south and east-west through the city. According to the seismologist Klaus Jacob, working at Columbia University, "an earthquake could occur anywhere."

There have been major earthquakes in the area of New York City in the past two hundred years. One happened in 1737 and another in 1884. Since 1884, there have been no *major* earthquakes in the vicinity of New York City. A quick calculation shows that the chance of a third earthquake taking place in the next thirty years is quite probable.

Part of the reason is that a level five earthquake isn't much news in California. Buildings are spread out more than the buildings of New York City, and the population isn't as dense. Since earthquakes are much more common in California and Alaska, eighty percent of earthquake hazard reduction funds are spent in those two states. Few people, including city officials, worry about a New York City earthquake. This means that Manhattan isn't really prepared if a major quake hits.

The infrastructure of Manhattan is much older than that of Californian cities, and it's much more vulnerable to an earthquake. In New York, a level five earthquake would be like a level six earthquake in California. Manhattan is set on bedrock. Bedrock might be unshakable, but seismic waves travel much faster in bedrock than in several underground plates. Worse, much of New York is built on soft soil. The Manhattan waterfront was built on reclaimed land and much of it is made up of sandy and loose rock. In the event of an earthquake, soft soil and landfill is shaken enough to turn it into fluid. A good part of New York City would be washed out to the ocean by a major earthquake.

An earthquake in New York would smash unreinforced masonry, destroying the famous New York brownstones and walk-ups. Pipes beneath the street would crack and explode. Stairways would crack and elevators would be knocked off their cables. Rooftop water tanks would fall, as would decorations like parapets and gargoyles.

Manhattan has a network of old gas lines and nineteenth-century sewers and cast-iron pipes. Water pipes beneath the street are particularly vulnerable to a quake, which could result in a loss of water pressure. Firemen would thus have a difficult time putting out major fires. Flight would not be an option, as bridges and tunnels would be down or blocked with debris.

"Cataclysm" is merely a comic book story. In graphic detail, it shows disaster striking at the heart of a great urban metropolis. In light of the attack on the World Trade Center, it paints a vivid picture of not only what could happen but what did happen in New York. And in the end, it shows how the city survives because of the efforts of heroes. Just like real life.

Chapter 4

# Under the Sea

## Aquaman and Sub-Mariner

### Undersea Heroes

**Marvel Comics were** started by pulp magazine publisher Martin Goodman in 1939. At the time, Goodman was publisher of the Red Circle chain of pulps, which included titles such as *Marvel Science Stories*, *Sky Devils*, *Best Western*, and *Complete Detective*. Goodman was noted for his sharp business sense and his ability to stretch a dollar. According to the comic book historian Les Daniels, salesman Frank Torpey convinced Goodman that comic books, not pulps, were the future of inexpensive publishing.[15] Torpey was working at the time with a group of writers and artists, packaging complete comic books that he then sold to publishers. Thus, Goodman was able to enter the comic book field without any staff or major investment. During the same period, the government was after Goodman for reprinting previously published stories yet claiming they were new material. When comics proved as lucrative as Torpey claimed, Goodman quickly dropped out of pulp magazine field and concentrated on comics.

The first Marvel superhero was Sub-Mariner by Bill Everett. The success of Superman had promoted a wave of new comic book

[15]Daniels, Les, *Marvel: Five Fabulous Decades of the World's Greatest Comics*, Abrams, New York, 1991, p. 23.

publishers, all with their own, slightly different super-powered character. The Sub-Mariner was originally designed by Everett for *Motion Pictures Funnies Weekly*, a failed promotional giveaway. Instead, he appeared in *Marvel Comics* #1, published in October 1939. In an interesting turnabout, the same comic also featured the debut of the Human Torch, a flame-based character. Both characters were immediate hits. Thus, water and fire built the Marvel empire.

As described by Everett, Prince Namor the Sub-Mariner was the teenage son of an American navy officer and a princess of the undersea kingdom of Atlantis, located beneath the Antarctic ice. A child of two worlds, Namor possessed powers that went far beyond either heritage. Not only could he breathe on the surface or underwater, but he also possessed super strength and could fly. Namor fought to protect his undersea civilization from the ravages of unthinking men. He was a hero and a villain combined, as he carried out one violent attack after another on the world above the water.

The Sub-Mariner proved to be a huge success. He was the first comic book character who wasn't wholesome and true-blue. Prince Namor was radically different from his rivals at DC. Often, Namor would make jokes and sarcastic remarks while destroying huge buildings. His rebellious nature appealed to teenage readers. The Sub-Mariner was a superbeing with an attitude.

Of course, as the war in Europe continued to worsen, Prince Namor began fighting the Nazis. The Sub-Mariner didn't like humans, but he hated the Axis powers. Interestingly enough, though the United States didn't enter the war until December 7, 1941, Sub-Mariner and other Marvel heroes started fighting the Germans many months earlier.

Seeing the popularity of Marvel's aquatic warrior, DC soon entered the underwater Olympics with a water wizard of their own, Aquaman. Aquaman first appeared in the November 1941 issue of *More Fun Comics*. The creation of DC editor and oftentimes writer Mort Weisinger, Aquaman battled swastika-wearing villains, who were never once identified as Nazis, instead being labeled as "submarine raiders." In the story, Aquaman explains his motives in typical DC style. "There is much evil in this upper world. I will do my share in punishing it."[16]

As to his origin and ability to breathe under water, Aquaman tells the survivors of a U-boat attack that his father was a world-famous scientist who, after the death of his wife, devoted himself to the study of the ocean's mysteries. This unnamed scientist discovered the remains of a lost undersea kingdom that he thought was Atlantis. Living in an underwater laboratory, the scientist was able to study the records and journals of the undersea people. Using scientific knowledge far beyond anything ever discovered above the sea, the scientist was able to teach his son "to live under the ocean, drawing oxygen from the water and using all the power of the sea to make me wonderfully strong and swift."[17]

It was Aquaman's father who also gave him the name "Aquaman." When his father died, Aquaman continued his studies in the underwater city, until events on the surface forced him to return to the world above the waves. It wasn't until much later in the series, in the mid-1950s, that readers learned that Aquaman's mother had been a princess of Atlantis, hidden beneath the sea, and that his powers were inherited from her.

Bill Everett's Sub-Mariner was a superhero with a capital S. Not only could he breathe underwater, but he also possessed super strength and the ability to fly. Mort Weisinger's Aquaman wasn't nearly as powerful. The cover of the first issue of *Sub-Mariner* (Summer 1941) showed the Marvel superhero upending a Nazi submarine. In Aquaman's first adventure, the underwater protagonist had his hands full battling several Nazi sailors. However, both characters were able to communicate with and command fish, dolphins, and whales. With their underwater allies, Sub-Mariner and Aquaman gave a new meaning to "sea power."

Sub-Mariner was one of the pillars of Marvel's original superhero stadium. Not so with Aquaman, who remained a second-string

---

[16]*More Fun Comics* #73, "Aquaman," November 1941, p. 4.

[17]Ibid., p. 4.

character in the DC lineup. In the 1950s, Sub-Mariner sank beneath the seas as the Marvel comic line drifted away from superheroes. It wasn't until the Silver Age of comics, in the early 1960s, that the character returned, this time as a villain in the pages of *The Fantastic Four*. Both he and Aquaman were rewarded with their own comic title during the huge upswing of interest in comics in the late 1960s. Though interest in the two fishmen has once again declined over the past two decades, both remain important supporting characters in their respective comic-book universes.

Despite their status as second-tier superheroes, Aquaman and Sub-Mariner have lasted for sixty years as comic book characters. Neither possesses a hardcore fan following like Superman or Batman, Spider-Man, or the X-Men. Yet, while hundreds of other superheroes have floated off to the Sargasso Sea of lost characters, these two have withstood the test of time. Why? The answer's obvious. Man has always been fascinated by life beneath the waves. Anyone who's ever stared at the surface of an ocean or a lake can't help but wonder what it would be like to breathe underwater. Inspired by man's eternal curiosity, Sub-Mariner and Aquaman endure as lasting testaments to humanity's fascination with the Earth's last unexplored wilderness, the vast lands beneath the sea.

## Our Aquatic Ancestors?

Any investigation of Sub-Mariner or Aquaman has to start with the lost continent of Atlantis. Not only are Sub-Mariner and Aquaman aquatic superheroes, but both of them are the sons of Atlantean mothers who dwelled in the ocean and married surface dwellers. Why Atlantis? How did a race of humans come to live under the sea?

Atlantis was first mentioned by the Greek philosopher Plato, in his dialogue[18] *Timaios*, written about 355 B.C. The continuation of

[18] A written discussion in which the characters, most notably Plato's teacher Socrates, lecture each other.

*Timaios*, entitled *Kritias* (each dialogue being named after the main character of the piece), was begun by Plato sometime afterward, but it was never completed. As far as can be determined, these two dialogues were the source of all the stories about Atlantis that followed.

In the dialogues, Plato describes how Kritias explains that, a hundred and fifty years earlier, the famous Athenian statesman Solon learned about Atlantis during a visit to Egypt. When Solon returned to Greece, he related the story to his brother, Dropides, who was the great-grandfather of Kritias and who passed the tale down to his descendents.

This dialogue within a dialogue described a great war between an earlier version of Athens, existing 9,000 years before Solon's lifetime, and the island empire of Atlantis. According to the story, a great earthquake caused Atlantis to sink beneath the waters of the Atlantic Ocean. Needless to say, the dialogue had nothing to say about super science, an advanced civilization, life beneath the ocean, or much else other than Plato's theories about government and the state. In fact, careful reading of the dialogues leaves it unclear if Plato meant for the Atlantis story to be taken as anything more than a carefully constructed example of his ideas in action. The philosopher offered no proof of the island's existence. But like many stories, once started, the tale of Atlantis took on a life of its own.

The history of Atlantis has nothing to do with science and actually very little to do with archeology.[19] Whether the original dialogue is based on myth, legend, fact, or whether it was constructed entirely by Plato to make a point no longer matters. Over the course of two thousand years, the Atlantis story has grown into an entire mythology, the source of thousands of metaphysical studies, religious tracts, novels, movies, television spectaculars, and the underlying theme of numerous comic book stories.

However, it was primarily in comic books that the actual concept of humans breathing beneath the sea was popularized (if not

---

[19]For those interested in Atlantis, we highly recommend *Lost Continents* by L. Sprague de Camp, Gnome Press, 1955 (reprinted many times), as a definitive work on the subject.

created). The notion that the inhabitants of Atlantis somehow adjusted to life under the water was never an integral part of the Atlantis mythos. The golden age of Atlantis fiction spanned the last few decades of the nineteenth century through the first thirty years of the twentieth century. During that time, dozens of mainstream novels were published about the lost continent and its discovery in modern times. However, the idea of water dwellers never arose. The inhabitants of Atlantis always turned out to be living in a glass-domed city (a popular theme) or wandering the world as immortals, still alive thousands of years after the sinking of their home city. Fishmen were not considered proper characters for "serious" novels.

Perhaps the earliest story to even suggest the possibility of genetically manipulating people and animals into hybrids was H. G. Wells's *The Island of Doctor Moreau*. However, the first major work to deal with developing different species of mankind for different environments was Olaf Stapledon's *Last and First Men*.

It's possible that one or both of these books inspired Bill Everett. Mort Weisinger, a science-fiction collector and fan, probably read the two novels. Another possible influence on the two creators was L. Sprague de Camp's pulp story "The Merman," which was the cover attraction for the December 1938 issues of *Astounding Science Fiction*. It's equally probable, however, that both heroes were merely Superman in the water, with Atlantis serving as a convenient explanation for their powers.

Despite twenty centuries of writing, searching, exploring, and analyzing, there's no more solid evidence proving the existence of Atlantis than there was when Plato first described it in classical Greece. Belief in Atlantis seems to require the same leap of faith asked by most religions. However, the notion of humans breathing underwater or at least spending much of their existence in water has taken on new life in the past twenty years with a fascinating, though controversial, theory regarding early evolution. Called the aquatic ape theory, it presents an interesting proposition that perhaps early man spent as much time in the water as he did on land. First discussed in a 1960 article by Alister Hardy titled "Was Man More Aquatic in the

Past?", this unusual theory has recently come under focus again due to advanced research techniques that study early fossils.

The evolution of man took millions of years. Remaining close to water has always been extremely important to survival. It was a guiding force during man's evolution.

During the Ice Age much of the water in the world was bound into glaciers, thus causing sea levels to be much lower than they are now. Land bridges existed from Korea to Japan, from Siberia to Alaska, from Sicily to Tunisia. It's likely that early man crossed all of these land bridges, spreading humanity to every region on Earth that it was possible to reach by foot.

However, some areas were never connected to other countries by land bridges. While Java was connected to Asia by a land bridge, the Indonesian island of Flores never was. Yet, archeological digs on the Flores have turned up primitive stone tools and fossils dating back 900,000 years. There's no evidence that early men knew how to make boats, so they either swam across the fourteen-mile channel separating the island from the mainland or floated there on tree trunks. Other similar examples of early man taking to the water to settle new territory exist. Very early humans were evidently quite comfortable with water.

Which leads us to the aquatic ape theory, proposed approximately forty years ago by Alister Hardy. The basic concept of the theory is that humanity developed some of its distinctive features in an aquatic environment and that very early humans spent much more time in the water than we do now.

Needless to say, this unconventional proposal wasn't well accepted among researchers of the time. One of the main reasons the proposal was considered outlandish was that it contradicted a much more popular theory known as the savannah hypothesis.

For seventy years, between 1925 and 1995, scientists believed that the savannah hypothesis explained how humanity evolved from an anthropoid ape to an erect hominid (human). This theory declared that apes living on the grassy plains of southern Africa, where water was scarce and competition for food was high due to the open nature of the plain, evolved into intelligent beings.

This view, that apes evolved on savannahs, was upheld by many prominent researchers and scientists of the twentieth century. The spread of the savannah in Africa was cited repeatedly as one of the prime motivating forces behind apes evolving into mankind. As for why apes walk on two legs, "uprightness gave its possessors a chance to see over the tall grass and watch out for predatory enemies like the lions and sabre-toothed big cats."[20]

However, major fossil discoveries of early ancestors of humanity made in the late 1980s and early 1990s contradicted the main premise of the savannah hypothesis. Examples of earliest man, such as "Lucy," were discovered in regions that evidently had been forestland, not savannah, millions of years ago. New studies about life on the savannah also wreaked havoc on the savannah hypothesis. Research showed that humans, even millions of years ago, wouldn't have been capable of living on the African plains. Scientists who had been so sure of the savannah hypothesis were no longer certain.

Using biological, anatomical, and physiological data collected by Dr. Marc Verhaegen, Elaine Morgan demonstrated other major problems with the savannah hypothesis. Morgan pointed out that modern humans lack sun-reflecting fur and are comparatively hairless. The cooling system of our bodies doesn't work well in hot, dry, exposed environments like a savannah. We have too many sweat glands, and we waste both water and sodium. We can't drink a lot of water at one time, another necessity for savannah-based life.

The rejection of the savannah hypothesis in 1995 by a number of leading researchers involved in the study of the development of primitive man gave the aquatic ape theory (AAT) new life. The premise no longer contradicted one of the basic facts of human evolution. In fact, much of what the AAT stated seemed quite possible based on new discoveries made by scientists studying early hominid fossils.

Evolution states that humans, gorillas, and chimpanzees share a common ancestor. However, while gorillas and chimpanzees possess

---

[20]Tobias, Phillip V., "Water and Human Evolution," *Dispatches About Human Evolution*, December 1998, p. 5.

a number of common traits, humans are much different from either of them. Somehow, we evolved on an entirely different path. The savannah hypothesis was long offered as an explanation for our human features. However, with the repudiation of that theory, we're left with a major question mark about how humans evolved.

The aquatic ape theory suggests that millions of years ago humans and apes diverged along different evolutionary paths due to a large group of them being forced to live for several million years in a flooded, semi-aquatic habitat. Though at first this theory sounds bizarre, there's a great deal of evidence supporting it.

For example, in his book *The Driving Force: Food, Evolution and the Future* (1989), Dr. Michael Crawford discussed the importance of water in human evolution. One fact stressed in the work is that "to develop the large brain characteristic of the hominids, a chemical known as DHA was necessary."[21] The lack of DHA in the food available on the savannah was another nail in the coffin containing the savannah hypothesis. However, DHA is abundant in the marine food chain, which seems to indicate that early humans spent much of their time in or near the water.

Humans are classed as primates, the order of animals that includes apes and monkeys. However, among the hundreds of different primate species, only humans are naked. There are two types of habitats that produce naked mammals—underground ones and water ones. Every other mammal with little or no fur lives in the water, like whales and dolphins, or spends as much time in water or mud as possible, like pigs and elephants.

Another important fact about mammals is that while fur provides fine insulation for land-based mammals, the best insulation in water is a layer of fat. Humans are the fattest of all the primates. We have ten times as many fat cells in our bodies as is normal for a primate our size. Again, there are two types of animals that have large amounts of fat, those that hibernate and those that live in water. Fat in humans, as in aquatic animals, is mostly deposited under the skin

---

[21]Tobias, p. 8.

instead of being stored internally as it is in land-based mammals. It's unlikely that this fat layer developed when mankind lived on the plains, as fat would slow down a hunting species. It's much more likely that this fat developed during a period when man was semi-aquatic.

Humans are the only mammals that walk on two legs. It's much more difficult walking upright than it is walking or running on four. Scientists once thought that early man first developed a large brain, and thus started making tools, before walking upright. According to this idea, humans started walking on two legs so they could use their hands to carry weapons. However, it's been established by fossil evidence that man walked on two legs before his brain grew and he made tools.

Proponents of the AAT argue that if primitive man's habitat was flooded, then men would have been forced to walk on their hind legs to keep their heads above water. Today, there are only two other species of primates that stand on the ground and sometimes walk erect. Both are monkeys that live in swampy areas or forests that are flooded every season.

Numerous other facets of human anatomy point to the possibility that mankind was once a semi-aquatic species. While few scientists are convinced that the aquatic ape theory explains everything about the evolution of early humanity, it's not easily ignored. The theory speculates that the first humans were a result of a major flood in Africa approximately five million years ago. The sudden change in habitat forced apes to evolve in a new direction, one that eventually led to today's people.

Taking the aquatic ape theory one step further leads us back to Sub-Mariner and Aquaman. What if some of those apes who learned to walk erect and developed larger brains adapted to their semi-aquatic homes? What if they developed gills and learned to live beneath the water? Could the AAT point not only to the development of modern man, but to modern fishmen as well? Perhaps Sub-Mariner and Aquaman aren't as impossible as they seem. Is there any way that humans could live under water instead of above it?

## Breathing Underwater

Aquaman and Sub-Mariner are half-human, half-Atlantean. As such, they've evidently inherited both the power to breathe air (like their fathers) and the power to breathe underwater (like their mothers). Without delving into the hows and whys of their parents' marital lives, let's take a look at one of the basic questions of natural science. Since water consists of atoms of hydrogen and oxygen, the two main components of air, and humans breathe air without any problem, why can't humans breathe underwater?

If only life were so easy. When hydrogen and oxygen gases combine to form water, two hydrogen atoms bond with one oxygen molecule. The resulting compound is nothing like the original elements. Moreover, fish don't breathe the oxygen that forms $H_2O$. Instead, they breathe the oxygen gas ($O_2$) that's dissolved in the water. Numerous gases dissolve in liquids, and fish aren't breaking down water into its two components but merely taking another form of oxygen to survive. For us to breathe underwater, we'd need lungs capable of doing the same work as gills—pulling oxygen gas directly from the water. There are several huge problems that make this idea nearly impossible.

Our lungs developed over millions of years of evolution to work with air. As such, the lining of our lungs can't separate the oxygen gas contained in water. Fish absorb the oxygen gas contained in water by using their gills. However, that process is a lot more complicated than mere breathing. One major stumbling block is that a container of water contains only five percent the amount of oxygen gas available in the same container filled with air. To get the necessary amount of oxygen needed to survive underwater, we'd have to breathe in twenty times more water than air. That would take an immense amount of effort. Besides which, water is much heavier than air, putting a greater strain on our lungs. Obtaining enough oxygen gas to keep a body running is impossible. The real problem is that human lungs just aren't created with anywhere near the surface area to absorb the oxygen necessary to keep our bodies functioning.

It takes a great deal more effort for a fish to absorb oxygen than for a person to breathe. Fish, however, are cold-blooded animals. They need much less oxygen to survive than warm-blooded creatures. Though they can spend much of their time underwater, dolphins and whales are warm-blooded, and they need lungs and air to live. Since none of the humans in the Aquaman or Sub-Mariner adventures ever comments about the cold, clammy nature of their skin, it seems fairly safe to assume that they are warm-blooded. This leads us to conclude that despite statements to the contrary, they are not strictly inhabitants of the sea, but do need to rise to the surface from time to time to grab a whiff of air. Which forces us to ask another question that's critical to life underwater. What about pressure?

## Pressure

If Aquaman and Sub-Mariner are warm-blooded and have lungs, then holding their breaths for a long time underwater is the least of their worries. Our two undersea adventurers are constantly shown diving to the ocean floor, which immediately raises the question of pressure. How do Aquaman and Sub-Mariner remain uncrushed at the bottom of the ocean? And equally interesting, how do they avoid decompression sickness, even death, every time they swim to the surface?

The "bends," the nickname given to decompression sickness, is one of the most dangerous problems of deep-sea diving. It wasn't until pilots experienced blackouts in World War II that the sickness was thoroughly studied, resulting in safety measures to prevent the extremely painful and sometimes life-threatening attacks. Unfortunately, despite everything we know about the bends, it remains a threat to our two underwater heroes.

Even the most relaxed people on Earth live under pressure: 14.7 pounds of pressure per square inch at sea level, to be exact. That's the weight of the air above us stretching to the boundaries of outer space. Scientists call that amount of pressure "one atmosphere."

When a swimmer goes underwater, the weight of the water above him (or her) adds to that pressure. At a depth of 10 meters (approximately 33 feet) beneath the surface, the amount of water above a swimmer weighs 14.7 pounds per square inch. Thus the pressure at this depth is already two atmospheres. The farther down a swimmer goes, the greater the pressure on his body. The only way to keep a diver's lungs from collapsing under the increased pressure is to inhale air at the same pressure as the surrounding water. The deeper a diver descends, the denser the air he needs to breathe. Air pressure is normally maintained using an air tank and a regulator. Exactly how Aquaman and Sub-Mariner accomplish this without any equipment isn't clear.

The major problem arising from these changes in pressure isn't from oxygen but from nitrogen. The longer a diver stays underwater and the deeper he goes, the more compressed air dissolves into his body tissue. If the swimmer rises too quickly to the surface, oxygen and carbon dioxide are easily reabsorbed into the bloodstream, but nitrogen is not. These tiny bubbles of nitrogen block small terminal blood vessels, causing intense pain throughout the body, especially at the joints. Dizziness, temporary paralysis, and even convulsions may also occur.

The bends can be avoided in several ways. One is for the diver to slowly rise to the surface, taking periodic breaks to allow the pressure in his body to readjust to his depth underwater. Another method of avoiding the bends is to breathe pure oxygen before and during the dive, thus eliminating nitrogen entirely from the bloodstream. Unfortunately for Aquaman and Sub-Mariner, neither solution seems practical. When a lifeboat is being shelled by hostile sea raiders, our heroes can't slowly swim to the surface. And neither man glides through the sea with an unlimited oxygen supply strapped to his back.

Humans can't breathe under the sea because we're not equipped with gills to separate oxygen gas from water. Even if we could, our lungs aren't big enough to pull out enough oxygen for us to survive. Compounding the problem is pressure. The deeper we go underwater, the greater the pressure on our bodies. The only way to survive the pressure is to breathe in air at high pressure, which, if we are not careful, leads to decompression sickness. All these factors combined seems to indicate that living underwater isn't possible. Sub-Mariner and Aquaman appear to be as unbelievable as most other superheroes. Yet, recent experiments suggest that we could be progressing towards a world where comic book science comes closer to reality.

## Fluid Breathing

Breakthroughs in science usually require breakthroughs in thinking. Humans can't survive underwater because of the lack of oxygen and the incredible pressure. Fish exist at incredible depths beneath the sea because they can separate oxygen from water and because the pressure inside their body equals that of the surrounding water. The important word is water. Replace it with fluid and suddenly the impossible seems merely improbable.

In the 1960s, Dr. J. Kylstra, a scientist at SUNY Buffalo, discovered that saline solutions could be saturated with oxygen gas at high pressure. Kylstra performed experiments on mice to see if they could move the saline solution in and out of their lungs while extracting enough oxygen to survive. The mice could breathe the liquid, but carbon dioxide built up too quickly in their systems. Still, it was the beginning of fluid breathing experiments.

Following Kylstra's lead in 1966 was Dr. Leland Clark. Clark substituted fluorocarbon liquids for saline solution, as both oxygen and carbon dioxide were both soluble in such liquids. Dr. Clark reasoned that animals could absorb the oxygen from the fluid and replace it with carbon dioxide. Like Dr. Kylstra, Clark worked with mice. His experiments were more successful, though carbon dioxide retention was still a problem. Dr. Clark did discover that the lower temperature of the fluorocarbons was directly related to how long the mice could survive in the fluid. The colder the fluid, the slower the mice breathed, which prevented a buildup of carbon dioxide.

Experiments using fluorocarbons continued throughout the 1990s with increasing success. Tests were performed on dogs without causing major damage to their systems. One of the problems working with fluorocarbons was that human tissues retained the fluorocarbons. The creation by Alliance Pharmaceuticals of perflubron, a fluorocarbon not absorbed by the body, finally made it possible to use liquid breathing in human medical procedures.

In 1996, doctors used liquid breathing to help premature babies born with lung problems. More recently, the technique helped doctors deal with patients having lung problems.

Research involving fluid breathing continues. It's hoped that someday this technique will enable divers to venture far beneath the ocean's surface without any risk of decompression sickness. Fluid breathing might even provide a unique rescue method for sailors trapped in disabled submarines. In James Cameron's movie *The Abyss,* characters use an advanced form of fluid breathing to dive to incredible depths. It's science fiction in the movie, but someday it might be scientific fact.

It's only one step beyond fluid breathing to breathing water. Given the tremendous leaps in human bioengineering, it's not beyond belief that humans may someday breathe water. The secrets of Sub-Mariner and Aquaman might be the stuff of comic books today, but in two hundred years, it could be reality. Maybe, just maybe, this is an example of the science of comic books becoming the science of real life.

## Talking to Fish

Throughout the adventures of Aquaman and Sub-Mariner, the two underwater superheroes talk to fish. Mostly, they command fish, though sometimes their finny allies bring them information or news about undersea problems. We're never actually told how this communication takes place—neither Aquaman or Sub-Mariner actually speaks to the fish underwater—but it's usually hinted that telepathy is involved. Nor do the fish ever disobey the commands given them by their half-human buddies—even when it means they have to put their lives on the line to save some incompetent humans.

How do fish communicate? Is there any method that Aquaman and Sub-Mariner could use to talk to them and get them to obey their every command?

There are twenty thousand types of fish. They make up the largest group of the vertebrates. Fish communicate in numerous ways. Much of their communication is done visually. Body language among fish includes certain movements, postures, body color, and color patterns. Many fish can change their color and do so as a part of mating or fighting. Many fish communicate by sound, grinding their teeth and rubbing their body parts together. Certain fish make sounds using the swim bladder, an air sac inside the body that regulates buoyancy. It's possible that Sub-Mariner or Aquaman communicate with fish using one or more of these methods, though how the fishmen change color is unknown. Most likely, our heroes use pheromones, chemicals produced by the body that can be detected by fish.

Pheromones carry specific messages that result in behavioral changes in other fish. These messages usually deal with food, enemies, or sex. Reactions to pheromones are automatic. Different species of fish create different pheromones, so communication between different types of fish is usually impossible.

The use of pheromones by insects and animals is a proven fact. Scientists are less sure that humans emit pheromones. It's been assumed for many years that humans don't, that we're beyond such primitive response systems. However, research over the past decade has suggested that people might produce pheromones in sweat. While no results have been proven beyond a shadow of a doubt, numerous companies have been quick to advertise "pheromone spray," guaranteed to attract attention from the opposite sex.

More research needs to be done on human pheromones before we can draw conclusions. Still, if we're capable of producing such odors, detectable only on a subconscious level, then it doesn't seem too outlandish that a hybrid being, half fish and half human, would be able to communicate with fish in the same manner. Since pheromones produce automatic behavior, sending signals in this manner would be tantamount to telling fish exactly what you want done without any worries that your command wouldn't be obeyed.

Like liquid breathing, commanding fish by pheromones is more dream than reality. Still, comic books are the stuff dreams are made of. With science progressing at breakneck speed, what seems impossible today could very well occur in the future. Aquaman and Sub-Mariner could be here sooner than we can imagine.

## Chapter 5

# Along Came a Spider

## Spider-Man

### With Great Power

**The most popular** character in the Marvel Universe almost never made it into comics. According to Stan Lee, when he first proposed a comic story featuring a man with the powers of a spider, Marvel publisher Martin Goodman indicated that he thought such a story wouldn't connect with comic readers.[23] Lee felt otherwise, and when one of Marvel's monster comics, *Amazing Adult Fantasy*, aka *Amazing Fantasy*, was slated for cancellation, Lee took advantage of the situation. He ran his Spider-Man origin story in the last issue of *Amazing Fantasy*. After its publication, Lee went back to work on other projects.

Months later, sales figures indicated that the Spider-Man issue of *Amazing Fantasy* was one of the best-selling Marvel comics ever published. Without realizing it, Lee had created another hit character. With a limited distribution contract, Marvel could only sell a certain number of titles each month. A place had to be found for Spider-Man. As mentioned in our chapter on the Incredible Hulk, that comic had to be canceled to make room for the new Spider-Man book.

[23]Daniels, Les, *Marvel: Five Fabulous Decades of the World's Greatest Comics*, Abrams, 1991, p. 95.

*The Amazing Spider-Man* #1 was published in March 1963. For a change, artist Jack Kirby didn't collaborate with Stan Lee. Instead, Steve Ditko, whose comic style was edgier and more angular than Kirby's, worked on Spider-Man. Ditko's people weren't any more or less believable than Kirby's people, but they looked a lot less heroic. It was a style that perfectly suited the mood of the comic. Without question, Ditko's work on the early issues of *Spider-Man* was a large factor in the series' incredible success.

The first Spider-Man story was a radical departure for comic books. It may have been the first story in the genre that recognized that superheroes were people and as such didn't always do the right thing. Or even the smart thing. That no matter how godlike their powers made them, their feet were firmly grounded in everyday life, and they had to deal with real-life situations.

Earlier comics had featured stories in which an out-of-luck character learned an important lesson from a superhero. *Spider-Man* was the first comic book in which the out-of-luck hero was constantly learning lessons from his own mistakes.

In the first Spider-Man adventure, bookish, wimpy Peter Parker visits a science exhibit where they are giving a demonstration about radioactivity. At the show, Peter is bitten by a spider that only moments before was accidentally bombarded by atomic radiation. On his trip home, Peter discovers the spider's bite has given him the powers of a spider. These powers include spider strength, spider speed, spider agility, spider grip, and what was labeled as his "spider-sense." We'll discuss these powers in detail later in this chapter.

Astounded by his new skills, Peter sews a costume, mostly to conceal his true identity from kids at school, and he builds two web-shooters, one for each hand, which will eject a sticky, weblike solution that hardens into silk ropes. Armed and protected from prying eyes, Peter goes out into the world to find fame and fortune. Unfortunately, all he discovers is a painful lesson in life.

During an audition for a show business job, Peter ignores police cries for help and lets a small-time crook get away. He feels stopping crooks isn't any of his concern. Peter reasons that's why there are cops. Then in an ironic and deadly twist, several days later, a criminal kills Peter's Uncle Ben, with whom Peter has lived since his parents died. At the end of the tale, Peter, dressed as Spider-Man, catches the crook and realizes it's the same one he let escape earlier. He realizes that if he had helped the police when he had a chance, his Uncle Ben would still be alive. Unlike most superheroes bent on a mission of revenge, Spider-Man's origin is based on responsibility.

Peter Parker vows to use his great powers for the forces of good. He quotes something his uncle told him: "With great power comes great responsibility." That motto works perfectly for Spider-Man, though it can be used for almost every superhero comic book.

Peter, as portrayed by Lee and Ditko, was the brightest and loneliest teenager in high school. He defined the word "geek" before it was invented. Peter had all the dreams and ambitions of a normal high-school teenager, but he was a magnet for trouble, a smart student lost among the jocks and glamour queens of his high school. In an effort to make the comic even more accessible to readers, Stan Lee often included thought balloons over complex emotional scenes so we could read Peter's thoughts every time he was taunted, pushed, or provoked at school.

*Spider-Man* was a tremendous hit with readers because it gave millions of teenagers a hero with whom they could identify. Jocks and cheerleaders didn't read comics. The quiet, introspective students did, the same type of students as Peter Parker. Spider-Man was the first ongoing, reality-based superhero comic. Peter could swing over buildings, use his artificial webbing to catch a pair of bank robbers, then return a missing animal to the zoo as part of a day's work. But in the evening, at home with his Aunt May, he was the one superhero who needed to sew the holes in his socks, take two aspirin for his headache, then watch the TV news to see if there were any reports about his exploits earlier in the day.

Peter was brilliant, dedicated, and hardworking, with a sly sense of humor. He just didn't know how to interact with girls, or, for that matter, with most teenage boys. The life of Spider-Man resembled the lives of his readers. It was the first comic book about teens that treated teens not as sidekicks but as heroes.

Spider-Man was a major success story for Marvel. Sales rose steadily until the title finally surpassed *The Fantastic Four* as Marvel's most popular comic. The continuing saga of Peter Parker's high-school life and times was as addictive as most TV soap operas. Plus, each month, Spider-Man fought some of the most bizarre and original villains ever to appear in comics. Dr. Octopus was a criminal genius with four long metal arms attached to his body, Sandman was a small-time crook whose body was completely made of sand, Vulture was a high-flying crook, and Lizard was an innocent research scientist who accidentally changed into a human-hating intelligent lizard.

*Spider-Man* remained Marvel's most popular title for many years. Steve Ditko, whose vision of the character differed from Stan Lee's, quit the comic in 1966, but the comic still remained incredibly popular, with artist John Romita taking over Ditko's position. *Spider-Man* was so popular that the switch in artists didn't bother fans.

After many years in high school, Peter Parker finally graduated and went to college. Again, while DC Comic characters never changed, Marvel's heroes grew and changed with the times. Marvel stories had continuity and DC stories didn't. By the late 1960s, Marvel was challenging DC as the most popular comic book line in America. Leading the charge was *Spider-Man*.

It was in *Spider-Man* that Marvel ran a three-part anti-drug story without the approval of the Comic Book Code. Lee felt the story was important enough to risk offending some corporate buyers. The gamble paid off, as Lee and *Spider-Man* got huge amounts of newspaper and media publicity for the story. Several years later, *Spider-Man* featured the first appearance of a deadly manhunter called the Punisher, who later became one of Marvel's most popular characters. In 1976, Spider-Man even got the chance to fight Superman, as Marvel and DC entered an unusual arrangement to allow their most popular characters to collide in the first-ever comic company crossover title. Spider-Man was so popular that he got his own cartoon series, his own live-action TV show, and his own newspaper comic-strip. In 2002, he starred in a multimillion-dollar action film.

Spider-Man's success was bolstered by interesting and often daring storylines. In perhaps the boldest move in comics up to that time, Peter Parker's college girlfriend, Gwen Stacey, was killed by the ongoing Spider-Man villain, the Green Goblin. Years later in 1987, Spider-Man married Mary Jane Watson, a character introduced more than twenty years earlier in the comic.

Perhaps the most controversial story ever to appear in Spider-Man started in *Spider-Man* #144, with the reappearance of Gwen Stacey, who had died two years earlier. A few issues later, in *Spider-Man* #147, it was revealed that the new Gwen was a clone of the original. Two issues later, Spider-Man fought and supposedly killed a Peter Parker clone. Or so it seemed.

Eventually, Peter married Mary Jane and life went on. Until mysteriously, Peter's parents, supposedly killed in a plane crash many years before, turned up alive. When it was revealed that the parents were fakes, Peter nearly went crazy and his elderly Aunt May suffered a stroke. It was at the hospital that Peter encountered his clone, Ben Reilly. But the question was soon raised whether Spider-Man was the real Peter Parker or the clone called Ben.

The clone saga ran from early 1994 through 1997, through three different monthly *Spider-Man* comics. The complex story, introducing one duplicate character after another, was so convoluted and contrived that it alienated many longtime Spider-Man readers. The plot, with its multiple Spider-Mans, manipulating masterminds, and explanations that one needed diagrams to follow, was finally resolved with the defeat of the secret supervillain and the death of all the clones, leaving the original Peter Parker free once again to pick up the mantle of the one-and-only Spider-Man.

Since his creation nearly forty years ago, Spider-Man has maintained his position as one of the most popular comic book characters ever created. He's perhaps Marvel Comics's most recognizable creation. Which leads us, as always, to ask, how believable is he? Could a man ever possess the powers of a spider, and what would they be? What's all this about clones? Is the clone saga a warning of things to come or just a wild story with no chance of ever happening?

## The Power of a Spider?

When Peter Parker was bitten by a radioactive spider, he acquired the human equivalent of a spider's powers. What exactly did that mean? According to early issues of *Spider-Man*, Peter gained strength, speed, agility, grip (the ability to walk on walls and ceilings), and what became known as his "spider-sense," a sort of ESP reaction that warned him about approaching danger. Anxious to capitalize on his new powers, Peter invented webslingers, small devices he wore under the sleeves of his costume that were finger activated and shot out a fluid that hardened into sticky silk cables that Peter could swing on as transportation or use to tie up the criminals he caught.

Stan Lee's superhero stories for Marvel Comics in the early 1960s revolutionized comics with their more realistic settings, their soap-opera style of storytelling, and their new and very different type of characters. As Marvel's editor and primary writer, Lee was one of the prime movers in the rebirth of superhero comics. He focused his stories not only on the superhuman aspects of his main characters, but also on their often troubled and difficult personal lives. Still, by concentrating his stories on the personalities of his protagonists, Lee sometimes didn't devote enough attention to the astonishing powers he created for his superheroes.

Lee's background was firmly rooted in the comic book field, while many of the creators working at DC had originally been science fiction writers and science fiction fans. As such, they were veterans of rationalizing amazing and sometimes outlandish stories by using fragments of real science. In the late 1950s, most of DC's new or revived Golden Age heroes were given origins that sounded plausible and that reflected in some small manner the science knowledge of the time.

That wasn't the case with Marvel's Silver Age characters. Personalities were the main concern and scientific accuracy never seemed to be addressed. Unfortunately, this lack of scientific relevance, as well as not maintaining a strictly logical viewpoint in the comics, often led to major inconsistencies in a hero's abilities. Some of Marvel's most fascinating and challenging characters were based on faulty (and sometimes totally inaccurate) beliefs about science. In these situations, inventing "new" science would have been better than crediting misguided science.

Thus in early Marvel comics, we had the Incredible Hulk created by atomic gamma rays—which were not only real but definitely not a Frankenstein's wand to create new life forms. In Spider-Man's case, the problem was even worse. Having Peter Parker bitten by a radioactive spider was a typical, very ordinary origin for a superhero. However, the powers acquired by Peter, supposedly those of a gigantic human spider, bore only a faint resemblance to those of real spiders.

The problem with Spider-Man isn't that he's improbable but that he's inaccurate. Let's briefly examine real spiders. Then we'll compare them to Spider-Man.

First and foremost, spiders aren't insects. Both are invertebrate animals (lacking backbones) and belong to the same phylum, or large group of living things, known as the arthropods. This group includes all invertebrate animals with jointed legs and a segmented body. Both insects and spiders are small in size and have exoskeletons made of chitin. They also both have a body cavity filled with blood that's pumped by a heart, and they have the same type of muscles and nervous system.

Insects have three body parts—a head, a thorax, an abdomen—and three pair of legs connected to the thorax. Spiders have eight legs and two body parts: a combined head and thorax known as the cephalothorax, and an abdomen. Insects have antennae, while spiders don't. Many insects have wings, but spiders never do. Plus, their breathing systems are very different.

There are over a million types of insects, but there are only 35,000 species of spiders. As far as can be determined by fossil records, spiders have existed in much the same manner for over four hundred million years. Spiders are known to us primarily for two things: their bites and their webs.

All spiders are predators, feeding on living insects, mites, and even other spiders. They capture and subdue their prey, which in most cases are insects,[24] by injecting them with poison from their fangs. Afterwards they cover them with silk webbing and devour them slowly.

While spider bites are a popular device in James Bond movies, they are rarely fatal to humans. A study done in the United States from 1989 to 1993 noted that less than 5,000 spider bites are reported per year. While this number still sounds rather frightening, it's hardly noticeable when compared to the 800,000 dog bites requiring stitches reported each year by the Centers for Disease Control. An Ohio State report emphasized that point by noting that collisions between automobiles and deer caused an average of 130 deaths per year; dog bites resulted in 20 deaths per year; and spider bites resulted in not one fatality in the entire four years.

All spiders spin silk, but many of them, such as tarantulas, jumping spiders, and wolf spiders, don't make webs. Silk is made by six silk glands beneath the spider's abdomen. It's a protein that's formed as a liquid and squeezed out of spinnerets very much as toothpaste is squeezed from a tube. After leaving the spinneret, the liquid thread hardens into silk threads. Spider silk is about one twenty-five-thousandth of an inch in diameter. Spiders can produce several types of silk, including one type for wrapping up prey it has caught, another stickier type for webs, and another type to make egg sacs. Some spiders even use silk threads as a sort of balloon or parachute to transport themselves from one location to another. Spiders have traveled incredible distances carried by air currents.

Spiders have long legs that are covered by hairs, spines, and bristles that are called setae. These setae are connected to the spider's brain and deliver all sorts of sensory information. Most of a spider's setae are located on its long, very flexible legs. All spiders have two claws at the end of each leg.

---

[24]A few very large spiders eat mice or very small birds.

Contrary to popular belief, all spiders *don't* have the ability to walk on ceilings or up the sides of bathtubs. Only hunting spiders have a thick group of hairs known as a scopula between the claws at the end of their legs. Each hair is made up of thousands of microscopic filaments that are covered with moisture and make it possible for them to stick to smooth surfaces. Web-spinning spiders don't have scopulae; instead they have an additional, third claw to grip their webs.

Now, what traits does Peter Parker, Spider-Man, share with the spiders we've described? A few, but not many.

Spider-Man's origin story of Peter Parker being bitten by a radioactive spider isn't a problem. In September 2000, radioactive insects were discovered near the Hanford nuclear complex in Richland, Washington. Radioactive red harvester ants were found underground near the complex. In August 2001, plant employees found radioactive flies in the plant's regular garbage cans. After checking the town dump where the garbage is taken, workers for the Energy Department found over 200 tons of garbage that had become radioactive due to contact with the bugs. No radioactive spiders were discovered, but in Spider-Man's origin story, the spider that bit Peter Parker died soon after infecting him. As far as we could determine, no one at the Hanford Nuclear site checked for spider remains.

Thus the radioactive spider wasn't impossible. Nor was the first spider power Peter acquired. Walking home from the science exhibit after being bitten, Peter was almost hit by a speeding automobile.[25] He leapt to the side and almost without thinking found himself climbing up the side of a building without effort. He soon discovered he could climb along a cable as easily "as the spider itself can glide along its web."[26]

Assuming Peter Parker gained the power of a hunting spider, his ability to walk on walls and stick to surfaces made perfect sense. We

---

[25]The driver of the auto remarks, "That was one egghead who won't daydream any more when he crosses a street." Evidently being smart in the 1960s put a target on your back.

[26]*Amazing Fantasy* #15, August 1962, p. 6.

should mention that Peter was shown using only his bare hands to climb the building, not his feet, which were covered with shoes. When he started wearing a costume, we can only assume that the material used for his gloves and boots was porous enough to allow fine hairs to pass through it.

As discussed earlier in this chapter, this wall-climbing ability was a skill possessed by hunting spiders. Peter's choice of words was rather strange, especially since he was a science major, but he evidently wasn't an expert in biology. Hunting spiders don't "glide along" webs, since they don't build webs. But then again, neither did Peter Parker. That's a skill he didn't possess.

However, the next spider power Peter acquired had nothing to do with spiders. When he reached the roof of the building he was climbing, Peter crushed a steel pipe without effort. Later in the same issue, he fought a professional wrestler and beat the man easily, using the speed, strength, and agility of a giant spider. To which we reply, *Huh?*

Spiders move quickly but are not noted as being particularly fast for their size. Anything with eight legs can move only so fast without falling over itself. A spider doesn't possess the coordination necessary to keep its legs untangled from each other at high speeds. More to the point, spiders have never been noted for their great strength. We can only assume that Stan Lee was thinking of ants and similar insects when he gave Peter super strength.[27] Gifting Peter with spectacular agility makes no more sense. Spiders aren't particularly agile. They scurry about on the ground or on their webs but are not noted for their grace and balance. Nor do spiders possess any great hunting skills. Actually, as noted in the book *Spiders: The Story of a Predator and Its Prey* by Dick Jones, "however clever they may appear in catching, for example, flies, spiders are in fact very inefficient predators. The web weavers do not actively catch their prey—the prey catches itself, rather like a fish caught in a net. Hunting spiders, particularly the nocturnal kinds, do little real hunting."[28] In

[27]We discuss ants and super strength in Chapter 8.

[28]Facts on File, 1986, p. 13.

nearly all circumstances, spiders avoid attacking anything approaching their size, as they are miserable fighters.

Thus, of Peter Parker's physical spider powers—spider strength, spider speed, spider agility, and spider grip, the only one that actually relates to a spider is the last, Peter's ability to climb and perch on any surface. Spider strength, spider speed, and spider agility have nothing to do with real spiders. What then of spider-sense, Peter's ability to sense danger before it strikes?

As mentioned earlier in this chapter, spiders are covered with hair, spines, and bristles, which are collectively known as setae. All setae have some sort of sensory function and they deliver information directly to the spider's brain. The setae at the end of spider's legs are extremely sensitive to numerous chemicals, enabling a spider to know instantly when it touches its prey whether the creature is edible. Another type of hair, known as trichobothria, can detect the slightest disturbance in the air made by the movement of possible prey. While neither of these senses involves detecting danger, endowing a human being with such finely tuned powers could result in a heightened feeling of awareness of his surroundings that would be the equivalent of Spider-Man's spider-sense.

Leaving us to conclude that two, spider grip and spider-sense, out of five of Peter Parker's super powers have any relationship with actual spiders. Obviously, Spider-Man's not really a spider-man. But then again, nothing about Superman's origins accounts for his X-ray vision.

We can't leave this discussion without a brief mention of Peter Parker's marvelous webshooters, the devices he invented to shoot incredibly strong silken ropes. In Spider-Man's origin story, Peter created his web shooters working with some odds and ends in his room at Aunt May and Uncle Ben's house. The invention of the devices took place in two panels. As Peter remarked a few panels further, "only a science major could have created a device like this." In the next panel, Peter mentioned that by using some "strong liquid cement at the end" of his web, he could use his webs to pull himself anywhere.

In a feature at the rear of *The Amazing Spider-Man* #2, we're told that Peter Parker is perhaps the world's greatest authority on webs and their creation. We also discover that Peter makes his web in a lab and keeps it in containers like "miniature toothpaste tubes." By adjusting the nozzle of his web shooters, Peter can spray his web as thinly as a rope, in a spray to form a web, or thickly, to create a strong adhesive.

Spider-Man's webslingers are Batman's silk rope pushed to the limits of technology, then beyond. Creating a small gun that attached under the wrist and fired a thin rope line would be a difficult job, but it could be done. Storing a supply of extremely high tensile strength silk beneath a costume would be equally difficult, but it could also be done. Unfortunately, that's where the *coulds* stop and the *could-nots* begin.

Spider-Man's webbing solution is beyond anything that could be invented by a genius science student in his bedroom. Or even working in his well-equipped high-school lab. There's no chemical compound on Earth that hardens from a liquid spray to a silken rope in an instant. Nor is there a substance that sticks to buildings and/or ceilings but not to hands. No substance exists that can be woven in seconds into a large netting that can be used to hold criminals captured for the police—and that flakes away when dry.

All these impossibilities lead us to an obvious question, one that's been asked again and again by die-hard Spider-Man fans for years. When Peter Parker was bitten by a radioactive spider, why didn't he inherit the ability to make silk organically? The writers at Marvel answered by simply stating that Peter didn't develop every spider trait.

Let's take a more in-depth look at the situation. Considering that Peter gained spider strength, spider agility, and spider speed, three powers not possessed by real spiders, why didn't he gain the ability to make webbing, a skill possessed by all spiders? If he had, then explaining his webbing wouldn't be difficult. Different spiders can make different types of silk, including extremely strong silk ropes, sticky or non-sticky webbing, and very sticky adhesive for trapping prey. If Spider-Man had been based on real spiders instead of perceived reality about spiders, he would have possessed the talent to make silk.

Of course, if Peter gained the ability to make silk, the immediate concern would be where would he make the silk, and even more important, where would it emerge from his body? Both questions are best not addressed, and perhaps in dodging them we answer the question of why Stan Lee preferred using webslingers.

What other spider talents should Peter Parker have gained when he was bitten? Endurance, perhaps? Spiders can exist for astonishing periods without any type of nourishment. In Dick Jones's book, *Spiders: The Story of a Predator and Its Prey*, the author describes keeping a spider in a test tube (with an airtight stopper) for a year, then releasing it unharmed. The creature lived the entire time without food, water, or fresh air.

Another power that would serve a superhero well is leaping. Jumping spiders are capable of leaping up to forty times their own body length when hunting prey. Assuming Peter is just under 6 feet tall, that would mean he could jump about 240 feet at a time. Maybe not as interesting a method of transportation as swinging on silk ropes, but a much more logical talent for a spider-man.

Still another type of spider, the fishing spider, is capable of running over the surface of quiet water. This type of spider can also crawl beneath the surface of water for up to thirty minutes. Water spiders, another species, live beneath the surface of lakes or ponds, breathing air they keep in a small bubble made out of webbing.

If Peter Parker was capable of making his own webs without resorting to mechanical aids and could leap over tall (or at least medium-size) buildings with a single bound, stay underwater for hours, and endure the most terrible tortures without suffering harm, then he would really be a spider-man. As defined by Marvel Comics, the best we can say about Peter is that he's a *pseudo*-spider-man.

## Clones, Clones, and More Clones

Cloning is one of the hottest science topics of the new millennium. It's been in the news for the past five years, ever since the first adult mammal, Dolly the sheep, was created by cloning. Though less publicized, animal cloning has continued, with ever-improving techniques. Pigs, mice, cows, and bulls have all been cloned successfully since Dolly was born. As we write this chapter, human cloning is being debated, banned, and planned, all around the world. Yet despite all the excitement, cloning isn't news to most science fiction fans or comic book readers.

Growing duplicate humans has been a staple in science fiction stories for nearly fifty years. In 1953, Poul Anderson's short novel *UN-Man* described a peacekeeping force of the future consisting entirely of the clones of one gifted man. More recently, Ira Levin's best-selling novel (later made into a film with Gregory Peck) *The Boys from Brazil* told of a secret band of German nationalists scheming to create another Hitler through cloning and behavior modification. Just a few years ago, the SF action film *The Sixth Day* dealt with the effects of cloning on future civilization and family bonds.

Cloning was natural for comics, which had been using the lunatic doppleganger theme for decades before science finally provided a logical explanation for evil twins. Clones and biological duplicates filled the world of comics. However, it was in the Marvel universe that cloning was taken to the limits of logic and beyond. The Spider-Man clone saga was perhaps the most controversial clone story published by Marvel in the company's history, but it was by no means the only one. Clones were prominently featured in Marvel comics for nearly thirty years.

*The X-Men*, Marvel's best-selling mutant team, featured clones in several major storylines over the decades. In the memorable "The Dark Phoenix Saga," Jean Grey, known as the Phoenix, died at the end of the story. Several years later, her boyfriend, Scott Summers, met Madelyne Prior, a beautiful young woman who bore a startling resemblance to Jean. Scott married Madelyne, and they had a son, Nathan. Then out of the blue, Jean Grey returned, not dead after all. In a major story arc known as "Inferno," Madelyne was revealed to be Jean's clone, grown by the evil Mr. Sinister. Madelyne wasn't even aware she was a clone until Sinister's disclosure, although the knowledge she was a duplicate person (i.e., a clone) and not an original (human being) helped turn her into a villainess called the Goblin Queen.

Several years later in another *X-Men* spinoff series, *New Mutants*, the son of Scott Summers and Madelyne Prior was revealed to be the mercenary known as Cable. Cable, who easily wins the prize as having the most convoluted history of any comic book character ever, returned as a tough old man from the future to the present day, to fight his clone, the menacing mastermind known as Stryfe.

Cloning, to the relief of just about everyone in the real world, hasn't come close to the amazing genetic magic performed in most comic books. While human cloning is going to be a hot topic for the next decade, it's doubtful that anyone expects to see clones similar to the ones who appear in comics for the foreseeable future.

The cloning that takes place in most horror novels, movies, and comic books offers a very one-sided viewpoint of the cloning debate. As has been pointed out time and again by doctors, the scientific process involved in cloning has nothing to do with creating some sort of fully grown, soulless monster. Cloning is about creating new life.

We're now at the threshold of cracking and using the *human genome*—that is, the body of genetic knowledge that defines who we are as individuals. While we have a long way to go before we can expertly use human genome information to manipulate our bodies and minds, clone ourselves, and eradicate diseases such as cancer and hemophilia, experts predict that major research will be under way in these areas tomorrow. And we literally do mean *tomorrow*.[29]

Put very simply, cloning is a fairly easy process to explain. All living creatures are made of cells. (For the sake of simplicity, we'll use a human male for our example, though a woman could be cloned in exactly the same manner. As could pigs, calves, sheep, and any other animal.) Our subject has different types of cells, such as blood cells, nerve cells, skin cells, etc., in his body. While each cell has a specific

---

[29]Gresh, Lois H., *TechnoLife 2020: A Day in the World of Tomorrow*, (ECW Press, 2001).

purpose, all cells in his body have the same DNA, the genetic code that contains all information for the cell to function correctly.

To clone our subject, scientists take some donor cells from the person. They then starve these adult cells (adult cells because they come from an adult) to force them into hibernation. This hibernation brings about a genetic reprogramming of the cell, switching the adult cells back to an embryonic state. Next, scientists take a normal unfertilized egg from a woman and remove the DNA from it. Taking the DNA-empty egg, scientists put it in a solution with the reprogrammed donor cell. An electric current is passed through the donor cell and the egg cell, fusing them together. The new cell is then implanted into a surrogate mother. The baby that results is genetically identical to the donor.

Cloning results in the birth of an exact duplicate (or triplicate or any number of copies) of the donor. Though the clone would have some minor differences from the original person, such differences wouldn't be noticeable beyond a molecular level. Many researchers think of cloning as little more than producing a time-delayed identical twin of a person. A startling procedure, but definitely not the creation of a soulless monster. One of the most important facts about cloning is that the clone and the donor would be identical but separated by a gulf of years, and, just like human identical twins, clones would *not* have identical fingerprints.

Returning to the world of comic books, video games, and SF movies, would evil scientists and mad doctors be able to clone anyone they wanted to without the knowledge and cooperation of the donor? Yes. As noted, all that's necessary to clone someone is a sample of their cells. As most people know from innumerable voodoo spell stories, such material isn't difficult to obtain. Only people who live in gigantic plastic bubbles would be immune from cloning. Anyone who visits a doctor, a hairdresser, or a dentist could be cloned. However, the resulting clone would be a baby, not a fully grown adult duplicate down to the most exact detail, as is always the case in comics and horror movies.

There's no technique for somehow impressing a person's DNA code on an artificially raised human body (as was done in the movie

*The Sixth Day*). Besides, growing bodies without a DNA code, which defines that body, is a basic contradiction in terms. There's no known way to age a human from a baby to an adult in a matter of days or weeks instead of years. Nor does any method exist to copy the information and thinking patterns of one person's brain to another.

A human being is more than a genetic code. Heredity helps determine what a person is like, but environment plays an equally important role. Identical twins share identical genomes, as do clones. However, it's been demonstrated repeatedly that if identical twins are separated and raised in different environments, they will become quite different persons.

If these twins don't eat exactly the same food, they won't have the same weight or grow to the same height. If one is raised Catholic and the other Protestant, they won't share the same religious beliefs. Nor would they think the same. Chances are they wouldn't like the same television shows. The same holds true for clones. People—and clones are people, not just copies—aren't destined to be the same as anyone else because of their genetic code.

To kill another comic book (and popular fiction) cliché, cloning requires a living cell from a donor. When someone dies, his or her cells die as well. Cloning a dead person, especially one dead for a long time, can't be done. So, despite plans to the contrary, we won't see a clone of whoever was covered by the Shroud of Turin anytime soon.

In comic books, cloning serves primarily as a scientific name for copying people. That's impossible. When human clones are born, they'll be real people, not counterfeits. That might not be as exciting as stories of evil doppelgangers attacking Spider-Man or a lost love suddenly returning from the dead. But it's exciting enough.[30]

---

[30]If you're interested in cloning and the genetic modifications that will be available within the next twenty years, you'll enjoy *TechnoLife 2020*, written by Lois and now available in paperback from ECW Press.

Chapter 6

# Green Lanterns and Black Holes

## Magic, Science, and Two Green Lanterns

**The original Green** Lantern was a DC Golden Age hero created by artist Martin Nodell, with some scripting help from writer Bill Finger. Debuting in *All-American Comics* #16, July 1940, Green Lantern described both the series hero, Alan Scott, and the source of his power, an ancient magic lamp, "the green lantern," made from the metal of a meteor that fell in China. Each day, Scott touched his ring, made of the same metal, against the green lantern. This supplied him with magical powers. As with many heroes from the Golden Age of comics, Green Lantern soon acquired a funny sidekick, Doiby Dickles, and fought a variety of bizarre villains. A popular character, Green Lantern appeared in the pages of his own comic as well as in *All-American Comics* until being canceled in 1949.

Science in the original Green Lantern series was never a concern. Not so when the character was revived in the pages of *Showcase* #22, October 1959. Like the Flash, another Golden Age hero revived in the Silver Age of comics, the new Green Lantern was given an entirely new origin, this time based on science, not magic.

In his first adventure, test pilot Hal Jordan finds himself drawn to the wreck of a flying saucer where an alien named Abin Sur lies dying. Sur telepathically explains to Jordan that he is part of a special galactic patrol that relies on a Battery of Power, a Green Lantern, to help battle evil throughout the cosmos. A special power ring obtains a cosmic charge from the battery that lasts for twenty-four hours. Unfortunately, due to a flaw in the unique metal that powers the battery, the ring is useless against anything colored yellow.

Sur, who crash-landed on Earth after being blinded by yellow lights in the atmosphere surrounding the planet, gives the power ring and green lantern to Jordan, naming the test pilot his successor. The alien dies, and a new hero is born.

Over the years, the Green Lantern has been passed on from one Earthman to another. It's gone from Hal Jordan to John Stewart to Guy Gardner to Kyle Rayner. Some years ago, Hal Jordan returned to the DC universe as a villain and an enemy of the Green Lantern. Forty years of storytelling makes for strange bedfellows. Despite his numerous galactic disasters, or perhaps because of them, the Green Lantern remains one of DC's most popular comic book heroes some forty years later.

The first issue of *Green Lantern* introduced the blue-skinned cosmic Guardians of the planet Oa, and these Guardians supply the power rings and green lanterns used by their galactic patrol. By issue #8, the Green Lantern of Earth finds himself involved in adventures with Green Lanterns of other solar systems. While a number of the stories in *Green Lantern* featured Hal Jordan and his successors fighting super criminals and foreign agents, the comic sparkled with enemies from space, battles in the future, and a menacing renegade Green Lantern aptly named Sinestro.

As the series continued, a more complex background was developed for the Galactic Guardians of Oa. Their science grew more and more powerful, making them nearly omnipotent. Then, in 1992, their reason for creating the Green Lantern Corps was finally revealed.

The story, "Ganthet's Tale," written by popular science fiction author Larry Niven, was an interesting attempt to blend cosmic history with real science. Unfortunately, the adventure, which had a misguided scientist looping the end of the universe with the beginning, thus bringing "ravenous" entropy from the end of time to the beginnings of creation, made little sense. A billion years of history was wiped out before it could happen due to the unstoppable forces of entropy. The adventure demonstrated that even a talented science fiction writer couldn't take a highly illogical concept and make it seem logical.

Green Lantern wasn't an alien being from another planet coming to Earth to use his super powers in the fight for justice. Nor was he the victim of cosmic rays or the bite of a radioactive spider. His career wasn't rooted in a quest for revenge against the killers of his parents. Hal Jordan was chosen as Green Lantern because he was honest, and he was a man without fear—stringent standards in the 1960s, though probably considered somewhat less than rock solid in today's world.

## Wanted: An Unlimited Power Source

Using the power from the ring, Green Lantern was able to fly through walls, transmute matter (in *Showcase* #22 he turns a metal battering ram into water), and catch a falling airplane. Except for the ring's inability to affect anything yellow, it seemed capable of doing anything.

This raised an inevitable question that was never adequately answered in the comic book series: What powered the Green Lantern? What source of energy did the blue-skinned cosmic Guardians of Oa use to energize not only the power battery of Earth's Green Lantern, but also the power batteries for all of the Green Lantern Corps throughout the universe? In noting that the Green Lantern Corps patrols the entire universe, we frantically do the math and discover that each of the approximately 3,200 Green Lanterns is in charge of a hell of a lot of space. Considering what we know about the size and number of stars in the universe,[31] that would give each guardian somewhere in the neighborhood of $10^{19}$ stars to watch. For those with a less active imagination, we're talking about 10,000,000,000,000,000,000 stars, or ten billion billion stars per

[31]See Chapter 1, about *Superman*.

guardian. Being a member of the Green Lantern brigade was obviously a full-time job.

As the Guardians of Oa are presented as the most intelligent life forms in the galaxy, it's not much of a stretch to assume that they had invented a way to channel the massive energy needed for the power rings from the green lanterns. With our increasing dependency on wireless phones and internet hookups, wireless transmission of power seems quite likely in years to come. The only question that really needs to be explained is the source of the huge amounts of energy used by the Green Lantern Corps every day. Since the Guardians are a peaceful race, they wouldn't steal their energy. It would have to come from a source that otherwise would be unused by other galactic races. What might that unknown fountain of energy be? Most likely, it's a black hole. Extending our reasoning one step further, perhaps the power lanterns are powered by one of the greatest cosmic puzzles in the universe—the elusive white holes.

Black and white holes are probably the most discussed and least understood subject in astronomy today. Along with their sibling scientific discovery, wormholes, they've served as short cuts and power sources for a dozen or more science fiction TV shows and movies. Exactly what are these mysterious objects, do they really exist, and how do they tie into the Green Lantern (as well as most other superheroes in need of a cosmic energy boost)? It's all a matter of life and death—not of people, but of stars.

## The Life and Death of Stars

Gravity is the most dominant force in the universe. Isaac Newton realized this over three centuries ago when he wrote that gravity was the major force in the universe and that the gravitational force between two objects was directly related to their masses. Two hundred years later, Albert Einstein revised Newton's ideas and proved in his general theory of relativity that gravity was a property of curved spacetime.

Gravity creates stars and gravity destroys them. Without gravity, there wouldn't be black holes. And without black holes, there would be any Green Lanterns.

Stars are huge furnaces formed by the force of gravity and whose life span is measured in billions of years. To people on Earth, the sun seems unchanging because in terms of human measurement, it is. The sun was born five billion years ago and is expected to burn for at least another five billion years.

Gravity is a force that tries to pull things together. If unopposed, gravity would pull everything into one small ball. On Earth, gravity is counterbalanced by the electrical force of atoms and molecules. But Earth and everything on it is made up of fairly complex atoms. That's not the case with stars.

Stars are created over billions of years. In the vast reaches of outer space exist huge clouds of hydrogen gas, hydrogen being the most common element in the universe. Approximately one of every sixteen atoms is helium, the second most abundant gas in the universe. These clouds are trillions of miles across. The density of such clouds is about ten atoms per cubic centimeter, hardly noticeable to us in a world where a cubic centimeter of silicon holds a billion processors and a cubic centimeter of the air we breathe is filled with $3x10^{19}$ atoms.

Some galactic event (perhaps the movement of a spiral arm of the galaxy, the shock wave of an exploding sun, or some reason we've yet to discover) causes this interstellar gas cloud to contract. When the atoms are squeezed together enough, gravity takes over and causes the contraction to continue, pulling more and more hydrogen and helium atoms together. The more matter in a particular location, the greater the gravity around that place. The cosmic cloud continues to contract, pressing atoms closer. With the increased pressure and atomic density comes heat. Gravitational potential energy is converted into heat. The cloud of hydrogen gets hotter and hotter.

The force of gravity continues to pull in more hydrogen gas, and the sphere created becomes larger and heavier. It has become what is known as a protostar. At its center, the temperature continues to rise as gravity squeezes the atoms. The tremendous heat increases random motion in the hydrogen atoms, sending their protons and electrons racing in erratic courses, often colliding. When the temperature at the core reaches ten billion degrees, colliding protons start to bind together and form new atomic nuclei. This process, nuclear fusion, results in the creation of helium atoms from hydrogen. However, the mass of the helium nucleus created is slightly less than the mass of the individual particles. The missing mass is turned into energy, per Einstein's equation $E=mc^2$. This thermonuclear reaction creates so much energy at the center of the protostar that it counterbalances the gravitational pull tugging the hydrogen gas that is not at the core. The protostar stops shrinking and stabilizes in size. It's become a star.

Scientists evaluate the size of stars by using our sun as a measuring stick. The amount of matter in our sun is called "one solar mass." Thus a star containing five times as much matter as our sun would have the mass of five solar masses. As noted earlier in our Superman chapter, many stars in the galaxy are less than one solar mass. Huge stars contain forty to fifty solar masses. Such stars are extremely bright, as they burn with incredible ferocity.

The mass of a star determines its ultimate fate. Stars with approximately the mass of our sun and smaller are considered low-mass stars. These stars evolve slowly, steadily burning their hydrogen fuel, changing into helium. This process that started billions of years ago continues for billions more, until the supply of hydrogen is used up. Without fuel, the hydrogen burning stops. Without this burning, gravity takes over again and the star's inner core, primarily made up of helium created from the hydrogen fuel, becomes unstable. Gravity compresses the helium core more and more until temperatures surrounding the core go so high that they ignite the hydrogen in the outer shell of the star surrounding the core.

At the same time, the core continues to contract, raising the temperature of the helium atoms. The same thing that happened to hydrogen at the star's core now happens to the helium. The atoms become so energized by the heat, now reaching 100 million degrees, that helium burning starts. Now, however, the fusion that takes place forms new atoms that are the next step up from helium, carbon and oxygen.

With helium burning taking place at its core, once again the force of gravity compressing the inner core is neutralized, and the core stops contracting. Instead, the two thermonuclear reactions taking place inside the star exert so much pressure on the outer layers of the star that they cause it to swell gigantically. The star evolves into what is known as a red giant. In our solar system, the sun would become so huge that it would expand outward over 100 million miles from its present boundaries, swallowing the planets Mercury, Venus, and Earth.

Still, the process of helium burning can only go on so long. When all the helium is finally burned away, leaving the more stable atoms of carbon and oxygen, the star's core once again begins to shrink due to the force of gravity. In a relatively short time (when considering billions of years), the star ejects most of the outer gas layers surrounding it. With no fuel for the star, the core contracts as gravity still pulls the mass of the center closer together. Soon the gas of the star is held together so tightly that the atoms are once again torn to pieces. As the core of the star shrinks down to the size of a planet like Earth, the loose electrons are squeezed together extremely tightly. Any further contraction of these electrons is impossible due to the fact that two electrons can't inhabit the same space. This new force battling gravity is known as degenerate electron pressure.

The star core is as small as it can go. The degenerate electron pressure balances the gravitational force, leaving the core approximately 10,000 kilometers in diameter. This dead star weighs about 1.4 solar masses. It is so dense that a cubic inch of core material weighs a thousand tons. The white-hot surface of such a star burns with intense white light. The core of the star has become what is known as a white dwarf star.

White dwarf stars are plentiful in our galaxy. They are the end product of suns like our own that have burned out after many billions of years. The existence of white dwarf stars depends on degenerate electron pressure. That's the force that stops gravity from contracting the star any further. Degenerate electron pressure (DEG) was discovered by astrophysicist S. Chandrasekhar in the mid-1930s. However, he also discovered that degenerate electron pressure is not infinitely powerful. The astrophysicist calculated there was an upper limit to the matter the force can support. This upper limit, called the Chandrasekhar limit, is 1.4 solar masses. Thus, all white dwarf stars must be 1.4 solar masses or less. What occurs to a dying sun that is 2, 3 or 5 solar masses?

These high-mass suns link our discussion of the death of stars to the Green Lanterns.

## The Origin of Black Holes

High-mass stars like Spica weigh much more than our own sun. Because of this extra mass, they take much longer to die. A star like the sun burns off all the hydrogen, then all the helium, at its core, and then, after much gravitational compression, it becomes a white dwarf. However, when a high-mass star burns off all the helium, leaving carbon and oxygen, the tremendous weight of the star's outer layers exerts so much pressure on these two seemingly inert elements that their temperature rises to 700 million degrees. At that heat, carbon burning begins. When that fuel burns away, the pressure continues to raise temperatures at the core to over one billion degrees, and oxygen starts to burn off.

The burning of oxygen atoms creates silicon atoms. As the core continues to shrink due to the huge gravitational pressures of the star's outer later, the temperature of the core reaches five billion degrees and silicon burning begins. Silicon burning results in the creation of iron atoms. Iron, however, does not burn no matter how high the temperature. Thus, a star with a solar mass much greater than that of the sun reduces to a core of iron, surrounded by outer layers of thermonuclear burning atoms of hydrogen, helium, carbon, oxygen, and silicon.

The weight of these outer layers continues to raise the temperature and pressure on the iron core of the star. Finally, the pressure becomes so high that the electrons of the iron atoms are squeezed into the atoms' nuclei. There, the electrons combine with protons to form neutrons. As neutrons take up much less space than the electrons and nuclei, the core suddenly shrinks. This core collapse releases a huge amount of energy, an amount greater than all of the energy radiated from the star for its billion years of life. The star explodes into what is known as a supernova.

Supernovas are extremely rare. Scientists estimate that they occur only a few times a century in our galaxy (that's three or four supernovas out of two hundred billion stars!). The last supernova visible on Earth occurred in 1604, and for a few days it glowed more brightly than the rest of the stars in the sky. However, supernovas have been observed in modern times in other galaxies. In 1987, a blue supergiant star in the Large Magellanic Cloud galaxy exploded in a burst of light visible to the naked eye. Labeled SN1987A, this supernova was extremely important to astronomers, as it was the first supernova that they could study with modern astronomical instruments. SN1987A confirmed numerous predictions that astronomers had made regarding supernovas.

A supernova results when the iron core of a high-mass star collapses into neutrons. After the huge explosion, what remains is a small but incredibly dense core of neutron-rich material. Though gravitational pressure is immense, this core can't collapse any further due to degenerate neutron pressure. The stabilized core is known as a neutron star. The concept of neutron stars was first raised by Fritz Zwicky in 1934, but a majority of astronomers doubted their actual existence until they were confirmed by radio telescopes many years later.

Neutron stars have a diameter of approximately twenty miles. Their density is incredible. A square inch of material from a neutron star weighs more than fifty billion tons.

Neutron stars are so small they can't be detected by telescopes. Instead, they're found by powerful bursts of radiation in the form of radio waves or X-rays. These bursts of radiation result from the neutron star's rapid rotation and powerful magnetic field. Neutron stars that emit powerful radio waves are known as pulsars.

White dwarfs are the result of a star of less than 1.4 solar masses dying. That's because degenerate electron pressure can't combat a gravitational effect of greater than 1.4 stellar masses. A neutron star results when a star greater than 1.4 stellar masses collapses. But as with white dwarfs, there is an upper limit to how much stellar material degenerate neutron pressure can support. That limit is 2.5 stellar masses. What then of stars with masses of fifty solar masses? When they collapse, they leave too much stellar material to be stabilized by either degenerate electron or degenerate neutron pressure. No other atomic forces exist. Gravitational force on such dying stars cannot be held back. The weight of trillions of tons of burnt stellar material presses on the core as gravity draws the mass smaller and smaller. The core shrinks until it becomes what is known as a black hole.

Black holes aren't a new concept. As far back as the eighteenth century, there was talk about "dark stars." In 1783, scientist John Michell stated that a star with a radius 500 hundred times the size of the sun with the same density as Sol would have such a powerful gravitational field that even light couldn't escape. Ten years later, French mathematician and philosopher Pierre LaPlace mentioned in his astronomy book that a star with a diameter 250 times larger than our sun would be invisible because the escape velocity of the star would be so great that it would require a speed greater than light.

However, the concept of black holes or "frozen stars" wasn't fully developed until Albert Einstein published his general theory of relativity in 1916. Soon afterwards, German astronomer Karl Schwarzschild wrote to Einstein describing the gravitational fields of stars using Einstein's new equation. In particular, Schwarzschild described a spherical star not rotating and gravitationally collapsed. The theoretical point in space where gravity collapsed that he discovered by using those calculations became known as a Schwarzschild singularity.

After work by Robert Oppenheimer (the man in charge of the Manhattan Project), George Volkoff, and Hartland Snyder, approximately twenty years later, scientists started believing that black holes might actually exist in our universe. Oppenheimer's work concluded: "When all thermonuclear sources of energy are exhausted, a sufficiently heavy star will collapse. Unless [something can somehow] reduce the star's mass to the order of that of the sun, this contraction will continue indefinitely."[32] When a star contracts to the point where no light is emitted due to the star's gravity, that star has become a black hole.

When a collapsing star contracts to where its mass and density are so great that gravity affects space around the star, making it impossible for even light to escape, the star is said to have fallen inside its event horizon. The event horizon is the borderline of a black hole. Anything outside it is not a prisoner of the incredible gravity of the black hole. Anything that passes the event horizon going inward will never emerge because the gravity is so strong that the object is pulled to the center of the horizon, the hole itself.

The center of a black hole is called a singularity point, because it is the spot where the star has reached infinite density and infinite mass, and has curved infinitely in spacetime. It is a hole in the fabric of space and time itself. The name "black hole" was coined by John Archibald Wheeler in 1968.

Once an object crosses the event horizon surrounding a black hole, there's no turning back. Nothing can travel faster than the speed of light, and even light can't escape the event horizon. Gravitational forces pull objects closer to the singularity point until the object is squeezed by unimaginable power and inconceivable force. Nobody knows for sure what happens to an object that passes through a singularity point. Hollywood screenwriters have suggested that it's a doorway to another universe, an elevation to a perpetual state of grace, or a gateway to hell. We, of course, believe a different idea, one we discuss a few paragraphs further on.

Black holes have been the subject of numerous articles and stories since the discovery of neutron stars in the 1960s. While most astronomers thought they might exist, no one actually expected to

---

[32]Gribban, John, *Spacewarps*, Delacorte, 1983.

find one. Not until the brilliant physicist Stephen Hawking proved that not only were black holes predicted by the general theory of relativity, but for the general theory of relativity to be true, black holes *had* to exist. Once Hawking showed that black holes had to be out there, the search began in earnest.

Since black holes by name are obviously black and their cores are not very big, finding them in outer space seems like an impossible task. However, many stars in the sky are binaries—that is, two stars close to each other and whose gravities affect the rotation of each other. A star system that moves strangely, as if being affected by another nearby star, though no star can be seen, is believed to indicate the presence of a black hole.

Plus, by their very nature, black holes are greedy. These perpetual devourers are always gobbling bits of matter. A black hole near a star leaves a visible trace of gas residue connecting an observable star to a black spot in the sky. In the past decade, NASA scientists and distinguished researchers from around the world have located a number of black holes.

The first candidate for a black hole in our galaxy is known as Cygnus X-1. This is an X-ray source in the constellation of Cygnus that corresponds to a star with between twenty and thirty solar masses. From the way in which this massive star moves, it's being pulled gravitationally by an invisible companion of between nine and eleven solar masses. Since it can't be seen, this companion must be very small and compact. Combining this knowledge with several other factors makes Cygnus X-1 the best candidate for a black hole in our galaxy.

Since the discovery of Cygnus X-1, a number of other black holes have been located by scientists studying the universe. One such black hole has been located in the constellation Aquila, about 40,000 light-years from Earth. This black hole, known as GRS 1915+105, is surrounded by a disk of matter sucked from a nearby companion star. A supergiant black hole, labeled M87, is suspected to exist at the center of the galaxy. In recent years, some astronomers have argued that there is a massive black hole at the center of our galaxy, the Milky Way.

Now, how does all this relate to the Green Lanterns or the Guardians of Oa? Black holes are the ultimate pits in the universal jungle. Whatever falls beyond the star's event horizon travels swiftly down to the singularity and is crushed into nothingness. The singularity at the center point of a black hole is a maelstrom of energy. It exists in a space we can't describe. We can't see past the event horizon, and even if we could put a camera inside the event horizon, we couldn't get any pictures back, as they would require light to pass through the event horizon, which is impossible. Black holes are the ultimate energy sink. But, since they are black holes, the collected energy has nowhere to go—other than through the singularity at the center of the hole. Into, perhaps, another universe, as suggested by some scientists and some science fiction writers? Or somewhere else in our universe? No one knows for sure.

There are several theories about black holes and energy. One extremely popular theory in science fiction books and TV is that black holes serve as gateways to hyperspace, a universe outside reality that connects to another black hole somewhere else in the universe. Anything is possible with black holes, but Stephen Hawking doesn't think this possibility would do future travelers much good. "I'm sorry to disappoint prospective galactic tourists, but this scenario doesn't work; if you jump into a black hole, you will get torn apart and crushed out of existence. However, there is a sense in which the particles that make up your body do carry on into another universe. I don't know if it would be much consolation to someone being made into spaghetti in a black hole to know that his particles might survive."[33]

Another possibility is that the singularity might be connected to a white hole. A white hole is exactly what the name implies—the opposite of a black hole. A white hole is a point in spacetime that explodes energy into the universe. Energy goes into a black hole, while energy comes out of a white one.

The main problem with white holes is that we've yet to find one. They exist only in the minds of scientists. No one has found a white hole, and no one is sure what to look for. Black holes, which take matter and energy out of the universe, are black blotches in the sky. White holes, which add matter and energy to the universe, are light and energy

---

[33]Hawking, Stephen, *Reader's Companion*, Bantam, 1992, p. 86.

sources in a sea of light and energy sources. A number of researchers believe that white holes might exist only in another universe, one parallel to ours, and that white holes are energy receivers transmitting the energy collected by black holes from our universe to another.

The Guardians of Oa in Green Lantern are described as the oldest and most intelligent race in the universe. Their civilization began several hundred thousand years after the Big Bang took place, approximately fifteen billion years ago. They know everything about our universe and the various parallel universes that exist outside our own. After one of their own conducted an experiment that destroyed a billion years of time, the Guardians founded the Green Lantern Corps to protect and guard the inhabitants of the universe from evildoers. Knowing that their space patrol would need a continual source of energy to defeat their foes, the Guardians gave them rings and power batteries. What fuels the power batteries? According to more recent issues of *Green Lantern*, they're all fed energy from a central power battery on Oa.

It seems logical to us, based on the research described in this chapter, that this central power source is a white hole. It's even possible that each Green Lantern power battery is connected to a separate white hole, thus giving each interstellar patrolman powers beyond anything else in the universe.

The most powerful superheroes in the universe need the most powerful weapons to make sure that justice has the strength to destroy civilization's enemies. But, as usual, with such power comes great temptation. It's a question that has been discussed often in *Green Lantern* comics and one that's sure to be raised again. When the watchers are given supreme power, who will watch the watchers?

## Yellow Light

We're told in the origin story of Green Lantern that due to a flaw in the power battery, the power ring won't work on anything yellow. However, if, as we've discussed, the power battery channels the energy of a white hole, how can we justify this weakness? If the Guardians are the greatest scientists in the universe, does it make any sense that they've created a weapon against the forces of evil with such a major defect? Is there any scientific reason that Green Lanterns would be useless against all things yellow?

A prism breaks white light into a spectrum of colors. The three primary colors are red, blue, and green. Mixed together, these three colors will blend into white. Mixed with each other, they create all other colors. Most colors contain small proportions of all wavelengths of the visible spectrum. Yellow contains an equal mix of red and green. Thus, it seems odd that the green light of the power ring wouldn't work on yellow.

Most color that we see is merely light reflected from objects around us. These objects, like buildings or cars, don't produce light but instead emit it through a process called color subtraction. In color subtraction, an object is struck by white light and certain wavelengths of the spectrum are absorbed by the object. Other wavelengths are reflected by the object giving it color. A green leaf appears green because in sunlight the leaf reflects green and absorbs all the red and blue color from the light. The brightness and depth of the color are determined by the exact spectrum of the wavelengths of the reflected light.

Cyan (blue-green), yellow, and magenta (red-violet) are what are known as subtractive primaries, or the primary pigments. A mixture of blue and yellow pigments yields green, the only color not absorbed by one pigment or the other. A mixture of the three primary pigments produces black. Mixed with each other, they form all the colors, just as in the primary spectrum. Again, it seems odd that yellow is part of the green light of the power ring, and yet yellow objects are unaffected by the ring.

More to the point, Green Lantern is often shown powerless against yellow-colored objects. Yellow is a color, a specific wavelength of light. It's not a property of specific objects. If Green Lantern can't stop a missile painted yellow, why doesn't he merely conjure up a huge bucket of green paint and slosh it over the rocket? Since Green Lantern's power ring works at the speed of thought (dependent on chemical reactions in the brain, thus somewhat less than the speed of light, but still incredibly fast), he'd be able to repaint the rocket hurtling through space at a few thousand miles per hour in an instant. After the paint job, he'd be able to handle the runaway missile with ease.

A fairly childish answer, but an answer nonetheless to a fairly simple plot device. There's no reason that Green Lantern's powers wouldn't work on yellow objects. It's a handy plot device to ensure that Green Lantern isn't all-powerful. But, logically, it makes no sense. Unless we assume the problem's psychological, not physical.

We're told throughout the *Green Lantern* series that the Guardians of Oa are the most intelligent race in the universe. In comics, intelligence usually equates with being technologically skilled. Perhaps, in the case of the Guardians, it also means wisdom.

If the power batteries channel the energy of a white hole into the Green Lantern Corps power rings, it makes these 3,200 beings the most powerful group of heroes in the universe. As stated so eloquently by Lord Acton, "Power tends to corrupt and absolute power corrupts absolutely." The process of selecting a wearer of the Green Lantern has been revised several times in Green Lantern stories. It's not foolproof. Sometimes it leads to disaster.

In a perfect world, no one would abuse power. But in a perfect world, there would be no need for policemen. Giving a mere mortal cosmic power with no boundaries to its use could cause more problems than it solves. In detailing the initial rise and fall of Sinestro,[34] it's clear that there are no moral boundaries established for using the ring. The only restrictions are in the minds of the users. Which is why it makes perfect sense that the Guardians would inform their galactic patrol that the ring doesn't work on yellow objects. In that manner, they make it clear to the Green Lanterns that the rings are not all-powerful and provide themselves with a psychological fail-safe in an emergency.

The power rings and power batteries aren't helpless against yellow. There's no reason they should be. The problem exists only in the minds of the users. It's a psychic roadblock placed there to keep the Green Lantern Corps from considering themselves gods. Because no one, not even the Guardians of the universe, knows if a god will be good or evil.

---

[34]*Green Lantern*, issues 7–50.

Chapter 7

# Of Atoms, Ants, and Giants

## Ant Man and the Atom

### Ant Man

**Before Marvel Comics** returned to superheroes with the publication of *Spider-Man* and *The Fantastic Four* in 1961, they were known as the home of giant monsters. Comic books like *Tales to Astonish*, *Strange Tales*, and *Tales of Suspense* specialized in stories like "I Found Monstrom, the Dweller in the Black Swamp," "Zutak! The Thing That Shouldn't Exist!" and "Moomba Is Here!"

These creatures came in many forms, but they all shared one common trait—they were huge. The beasts combined the freedom of comic book illustration with the size and novelty of Japanese and American monster movies of the time. Thus, while humans were menaced by *Godzilla* and *Rodan*, *Them!* (giant ants), and *Tarantula* on the screen, they found themselves fighting the Abominable Snowman, Grogg, and Fin-Fang-Foom on the printed pages of Marvel Comics.

In late 1961 and early 1962, Marvel switched from giant monsters to superheroes. Gone were the Crawling Creature, Pildoor, and the Gorilla Man. They were replaced by Spider-Man, the Incredible Hulk, and the Fantastic Four. However, despite the change in style, Marvel didn't entirely abandon the gigantic and the small. They merely combined the two genres into one, giving readers a taste of both worlds in a single story.

Marvel's first size-changing hero was Ant Man. In his January 1962 origin story, "The Man in the Ant Hill" *(Tales to Astonish #27)*, Dr. Henry Pym discovered a serum that enabled him to shrink to the size of an ant. In his second adventure, "The Return of the Ant Man," published in *Tales to Astonish* #35, September 1962, Pym created a special costume to wear when he shrunk to insect size, along with a helmet that allowed him to communicate with ants.

While Ant Man was a fun character, there really wasn't much he could do against his normal-sized foes. Even the addition of a female companion, the Wasp, didn't help the series. After a year's worth of adventures, Dr. Pym learned how to use his serum to increase his size and became Giant Man. In his new guise, Pym battled such foes as the Top, the Android, and the Eraser.

Giant Man proved somewhat more popular than Ant Man, and he continued in his own series for several years. Along with the Wasp, whom he married, Pym helped form the superhero group the Avengers. Later, a smaller Giant Man changed his name to Goliath. Then he reduced himself back to bug size, in yet another identity change, as Yellowjacket. Years later, Pym gave his Ant Man serum and uniform to a friend, who became known as Ant Man II. For a brief time, Hawkeye, a member of the Avengers, assumed the identity of Goliath. Needless to say, a Marvel villain named Power Man also possessed Giant Man's size-changing power.

Giant men and ant men weren't concepts unique to comic books. Both types of heroes were common in popular fiction long before comic books were created. Jonathan Swift wrote about tiny and gigantic people in *Gulliver's Travels*. Alice experienced difficulties with both sizes in *Alice in Wonderland*. Author Ray Cummings specialized in stories about people shrinking to microscopic size as well as growing incredibly large. Cummings, who worked for years as Thomas Alva Edison's secretary, wrote numerous adventures such as "The Girl in the Golden Atom," "The People of the Golden Atom," and "Beyond the Vanishing Point." Long before Giant Man's first appearance, Donald Wandrei wrote "Colossus," the story of a man who grew bigger than the universe. Decades before Ant Man, Henry Hasse told the fascinating tale of "He Who Shrank," while Paul Ernst wrote "The Raid on the Termites."

Assuming Dr. Henry Pym actually existed and invented a serum that enabled the user to become very small or very large, what would be the scientific risks involved? Could a man or woman survive as Ant Man and the Wasp? Would a normal person be able to move mountains if he became a giant man? What risks are involved with each transformation? And what about those science fiction movies of the 1950s? Are we in risk of being overwhelmed by giant ants or man-eating grasshoppers? What's the truth about the very large and the incredibly small?

## The Square Cubed Law

One of the fundamental rules of science is the square cubed law. It was discovered by Galileo in the seventeenth century. In its most basic form, the square cubed law states that when you double the size of an organism, you increase the surface area by the square of two, that is, four times, and you increase the volume of the organism by the cube of two, or eight times.

As with many scientific principles, the square cubed law is easier understood by using an example. Assume we have a cube that's one-inch high, one inch wide, and one inch long.

Let's also assume that the weight of this cube is one pound. Multiplying the height times the width gives us a crosssectional surface area of a face of the cube as one square inch. We mention surface area, as strength is directly related to the cross-section area of muscle or bone.

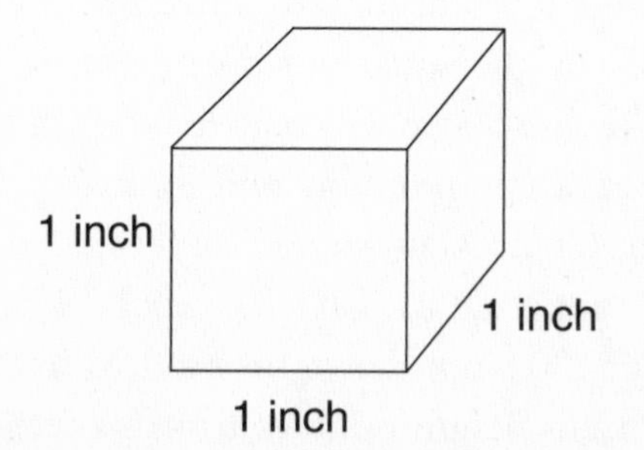

Now let's double the size of each side of the cube to two inches. This gives us a new cube looking like this:

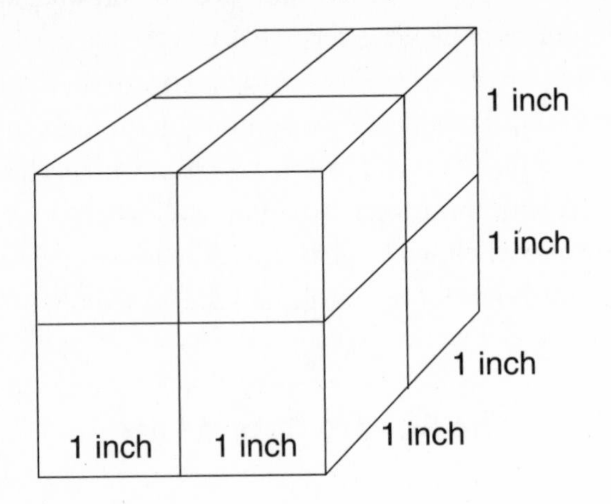


Our cube now is twice as long, twice as wide, and twice as high. However, the cross section of the face of the cube is now four square inches (2 inches × 2 inches), which implies that the strength of any cross-section of muscle or bone is four times as great as before. Also, in doubling our measurements, the volume of our cube has changed from one cube to eight cubes (2 × 2 × 2), and, therefore, the weight of our cube has gone from one pound to eight pounds. While the cube is twice as tall, it is four times as strong and eight times as heavy.

Both facts are extremely important to biomechanics. The square cubed law shows that if we double a man's size, his muscle power will increase four times, but he'll weigh eight times as much. Therefore, his strength will actually be cut in half. Since the strength of his leg bones increases only by the square of his height, while the weight increases by the cube, his bones will buckle beneath the pressure.

Now let's plug in a few numbers and look at our results. Let's assume Henry Pym, Giant Man, stands six feet tall and weighs 200 pounds. Henry, being a scientist, isn't incredibly strong, but he's no slouch, and he can lift 100 pounds over his head. When he takes a dose of his growth serum, his height doubles from six feet to twelve feet.

His weight goes from 200 to 1,600 pounds, while he's now strong enough to lift 400 pounds. Unfortunately, Henry can't stand up. Ankle and knee bones would shatter under such weight. The tallest humans on record stand about eight feet tall. A real Giant Man needs the legs of a Tyrannosaurus Rex (as well as its tail for balance!).

Let's push Dr. Henry Pym to the limit. According to the files on Giant Man, he's able to reach a height of sixty feet, or ten times his original size. Again, by the square cubed law, Henry would now weigh a thousand times his original weight ($10 \times 10 \times 10$), or 200,000 pounds. He'd be able to lift a hundred times what he lifted at his original size, or 10,000 pounds. More important, his bones would only be able to support 10,000 pounds, or one-twentieth of his weight. Poor Henry would be a gigantic protoplasmic slug, unable to move, with bones crushed to powder by his immense weight. Not a very heroic figure. But it's his own fault. Every scientist knows the square cubed law. Henry has no one to blame but himself for growing to such an unmanageable size.

Continuing in the same vein for a moment further, let's consider the stories and movies about giant bugs that were so popular in the 1950s. What would really happen to ants if they grew to the size of people?

First, we need to make a simple but important point about the differences between ants and people. Any comparison between humans and insects is fairly ridiculous. Ants and other insects are built entirely differently from humans. Our skeletons are inside our bodies, whereas theirs are on the outside. We have two arms and two legs, while ants have six legs and jaws that are used for lifting. Plus, our muscular structure is totally different. Ants are designed for lifting and we're not. We're not making excuses for humans, just setting the record straight before we continue our comparison.

Let's take an ordinary carpenter ant and increase it to human size. An ant is one-quarter inch long, weighs approximately 3 milligrams (we'll convert to pounds when the numbers become a lot bigger), and (as discussed in innumerable comic books and bad science fiction novels) can lift fifty times its own weight (around 150 milligrams). We'll feed it some of Dr. Pym's growth formula and increase it in size to approximately six feet long, about the size of the good doctor.[35] That's 296 times its original size; for the sake of simplicity, we'll round 296 to 300.

Using the square cubed law and running our calculations, our six-foot-long ant would weigh approximately 178 pounds. It would be able to lift 30 pounds. Not exactly the stuff of nightmares. Most likely our giant ant wouldn't even be able to lift itself off the ground. Sixty-foot-long grasshoppers? No problem.

The square cubed law works in both directions. If it's true for ants growing to the size of humans, then logic dictates that the law is true for humans shrinking to the size of ants. What would happen if a man like Dr. Pym were actually able to become Ant Man? Would he be as strong as his pals in the ant hill, or would he be as weak as a flea?

Let's again look at Dr. Pym, at six feet tall and 200 pounds. He still can lift 100 pounds. We'll have him shrink to the size of a carpenter ant, one-quarter inch long (or in this case, tall). Again, we're using a factor of 300, but now in the exact opposite direction. Performing the calculations again, we discover that Dr. Pym will weigh about 3.1 milligrams (just about the same weight as our carpenter ant). However, the same strength calculations determine that Dr. Pym can lift 500 milligrams. That's approximately 166 times his own weight, which is more than three times what a carpenter ant the same size could lift—proving perhaps that Dr. Pym would really be king of the hill.

## The Atom

Marvel Comics wasn't the only publisher who had a hero who could shrink to the size of an ant. The October 1961 tryout comic book, *Showcase* #34, featured the debut of a character known as the Atom, billed as "The World's Smallest Super-Hero." The comic was edited

---

[35]In *Them!*, the classic giant ant movie, the giant ants appeared to be about twelve feet long!

by Julius Schwartz, who was responsible for most of DC's Silver Age revivals of comic book characters from the Golden Age of comics. He named the hero of the Atom series after his old friend Ray Palmer, a well-known science fiction fan, science fiction magazine editor, and publisher of the first magazine about flying saucers. The name "the Atom" came from a 1940s DC character who had no super powers but was a short, tough, crime fighter who ran around wearing a cape and mask.

The Silver Age Atom as resurrected by Schwartz was a straight science fiction character, a hero who used super-science to develop a shrinking ray. The Atom appeared in two more issues of *Showcase* and was popular enough to move into his own comic book series, which lasted six years. The comic was canceled in 1968, but the saga of the Atom didn't end there. More than twenty-five years later, Ray Palmer returned as a character in a major DC comics event, "Zero Hour." In the story, thirty-year-old Ray had his aging process reversed, and he ended up being a seventeen-year-old with just foggy memories of his exploits as the Atom. Still, Palmer retained his shrinking powers, and at the end of the story, the young Ray joined the teenage superhero group the Teen Titans.

In his origin story in *Showcase* #34, the graduate student and fellowship research physicist Ray Palmer described how he saw a meteorite plunge to earth late one night. Digging deep into the ground at the crash site, he found what he identified as a piece of a white dwarf star. Ray carried the fragment of the star (a stone the size of a football) back to his lab, where he was working on a reducing ray. It was Ray's dream to compress matter without losing any of its physical and chemical properties. Thus, whole truckloads of crops could be shrunk to miniature size, carried from one part of the country to another as small packages, and then returned to normal size to feed hungry people. In some unexplained manner, Ray cut a piece from the white dwarf star fragment and made a reducing lens from it. When he passed ultraviolet light through this lens, the light ray shrunk inanimate objects. But these shrunken objects exploded after a short while due to the instability of the compressed atoms.

Later in the same story, Ray, his girlfriend, and a bunch of school kids were trapped by the collapse of a large cave. There was no way out except for a tiny hole no bigger than an ant could crawl through. Luckily, Ray was carrying his reducing lens with him. Ray placed the lens on a handy stalagmite and shrank himself down to ant size. Using the handy diamond ring he just happened to be wearing, Ray cut away rocks from the tiny hole, making the opening large enough for ordinary humans to crawl through. Afterward, Ray returned to normal size without exploding and carried everyone to safety. At the end of the story, Ray reasoned that some unknown force in his body enabled him to return to normal size.

In the entire story, Ray Palmer, the Atom, never once wore a special costume to disguise his features, nor did he get into a single fight. Needless to say, that didn't hold true for his second adventure, "Battle of the Tiny Titans," which appeared in the same issue.

The Atom's second adventure took place several days later. By then Ray Palmer had made himself a miniature costume using the material of the white dwarf meteorite. When he stretched the costume to fit his body, it became invisible and intangible. A control lever on this invisible uniform sent a wave of ultraviolet light through the costume, shrinking Ray's body and uniform. Though there was no mention of a size modifier, there obviously was one built into the mechanism, as Ray was able to shrink to whatever size was necessary in that adventure and those that followed.

In the course of his battle with an alien space traveler, Ray performed for the first time what was to be his most unusual stunt as the Atom. He dialed the villain's phone number, turned on a metronome, shrank to microscopic size, and then, when the phone was answered by the crook, traveled through the phone line to the other receiver! The amazing feat couldn't be explained in the panels of the comic, so a special footnote at the end of the issue gave the readers the details.

The explanation at the end of the story rationalized that once Ray dialed the correct phone number, he leapt into the mouthpiece of the telephone and shrank, "merging the atoms of his microscopically small body with those of the thin diaphragm which is part of the transmitter."[36] When the connection was made between Ray's phone and that of the villain, the metal diaphragm vibrated due to the sound waves of the metronome, which in turn vibrated against the carbon granules in the microphone. The vibrating granules sent electric signals through the metal wires to the other phone's receiver. The Atom, we were told, traveled along the telephone wire along with these electric impulses and was reassembled at the other end of the call.

By now, if you've been reading this book chapter by chapter, your brain should be screaming in pain. The Atom is a superhero whose origin and powers sound reasonable, but when examined under the harsh light of reality, they collapse into scientific techno-babble. Let's look at the Atom and see if he is any more believable than Ant Man. Or does a white star shrinking lens make any more sense than a shrinking potion?

## The Atom Exploded

Reading comic books requires some suspension of disbelief. When we pick up a comic, we need to be willing to give the author some leeway in his writing. Except in extreme cases, we have to be willing to accept some compromises. In comics, odd coincidences are allowed to help the plot move. Characters can change their minds suddenly, because, though most of us hate to admit it, we do the same thing. Situations are resolved in most books by the end of the story because readers like tidy conclusions.

Many classic science fiction stories are based on the premise What If? What if hostile aliens visited Earth? What if time travel were possible? One unlikely premise is made or an unproven conjecture is assumed for the sake of a good story. A good writer can take a single premise and construct an entire novel, or even a series

[36]*Showcase* #34, October 1961, p. 32.

of novels, based on one logical extrapolation of events or ideas. And then there are comic books.

Superhero comics expand the What If? concept into a premise for a hero or heroine. Batman, for example, is based on the premise of what if a young boy saw his parents gunned down by a crook and decided to dedicate his life to preventing crimes? Or, in the case of Spider-Man, what if a young man bitten by a radioactive spider gained spiderlike powers, and he used these unusual powers to fight crime? The concept supplies our character. The details make the character believable. Batman's arduous training and years of studying crime give him credibility as a character. Peter Parker's soul-searching monologues about his purpose in life and his spider-powers are what make Spider-Man convincing. Unfortunately, such characters are few and far between in comics. For instance, details destroy the Atom.

When we first meet Ray Palmer, we learn that he's a scientist (or a graduate student in science) investigating ways to compress atoms to help humanity. Immediately, we have our doubts about Ray, because everything we know about atoms says that they can't be reduced in size except in extreme conditions. There is space in atoms,[37] but quantum theory makes it quite clear that this space isn't empty. Electrons are zooming around the nucleus at incredible speeds. Atoms can be compressed, but only under conditions of extreme heat and pressure, like the inside of a collapsing sun. If you want more details, check out our Green Lantern chapter.

Another problem with Ray's story is his description of finding a piece of a white dwarf star that has fallen as a meteor to Earth. While a white dwarf star, the ancient remnant of a star that has burned away most of its gases, might sound harmless, it is anything *but* harmless. For one thing, the temperature on the surface of a white dwarf is still 20,000 degrees centigrade. Hopefully Ray wore insulated gloves when he picked up the fragment. Picking it up might have been a

---

[37]We also discuss this subject in our chapter about the Flash.

problem as well, since it's estimated that a square inch of matter from a white dwarf star would weigh *several tons*. The football-sized rock probably weighed 100,000 to 150,000 pounds, maybe more.

After Ray got the meteor back to his office and it hopefully cooled down, he took part of the fragment and cut and shaped it into a lens. These actions require more than suspension of disbelief. A star fragment is made up of gas compressed tightly by incredible pressure. That's not a material that can be shaped into a lens. Nor would the material work very well as part of a uniform to stretch over Ray's body. We've left the science fiction world of What If? and jumped headfirst into the land where facts aren't important when you're telling a story.

By the way, as mentioned in the Superman chapter, the light from a white dwarf star is still starlight. Shining ultraviolet light or sunlight through Ray's lens would still give us light, nothing more.

There are logic problems, as well, with the Atom's origin story and adventures. He wears a suit that's intangible and invisible but appears (in bright blue and red colors) when he shrinks to six inches tall. This leads us to surmise that the suit acts as the shrinking lens. If that's the case, how does Ray keep the suit not working in sunlight (which contains ultraviolet light)? Equally important, when the suit is exposed to the ultraviolet light, does it shrink along with Ray? The lens made from the white dwarf star that originally turned Ray into the Atom didn't shrink itself—it merely cast a light beam that made our hero shrink. If the suit is acting as the lens, wouldn't the suit remain unchanged? That's the implication in the Atom's adventures: that the suit has just been greatly stretched, and, when Ray's body shrinks, the suit material comes together and thus is visible.

The concept sounds logical until we start thinking about Ray's size. If the suit is visible and form-fitting when Ray is six inches tall, then what happens when Ray is only 0.5 inches tall? Does the suit still fit snugly? Or when Ray is 0.005 inches tall? Assuming the suit will fit Ray at all these sizes, how did he manage to stretch it to put on in the first place? Or even construct (dare we say sew?) such an incredible outfit? Science doesn't just require interesting concepts, it also requires logic. A suit that acts as a shrinking agent and adjusts to any size doesn't make much sense.

What about the telephone transportation trick? It required a print explanation in the letter column of the first issue of *Showcase* #34 to clarify how it worked. Does the explanation make any sense, or is it more technobabble?

As mentioned earlier, six-inch-tall Ray dials a phone number. (Since the story took place in 1961, poor Ray had to use a rotary dial. Push-button phones later made life easier for superheroes.) With a metronome in the background providing a steady noise, Ray shrinks smaller and smaller, and then leaps into the receiver. When the crook answers the phone, Ray completes the circuit between the two receivers. The Atom is sent speeding over the phone wires, and he jumps from the other receiver, much to the surprise of the criminal. According to the explanation, "The Atom was hurled along at telephonic speed by these electric impulses."

A transducer is a device that converts input energy to output energy. In a telephone, the transducer is a microphone that converts sound energy into electrical energy. The telephone microphone in the Atom stories is a carbon microphone. Beneath the mouthpiece of the phone is a vibrating diaphragm that is mounted in a cup filled with carbon granules. An electric current from a circuit flows through the granules. When someone speaks into the phone, the sound waves vibrate the diaphragm, which press on the carbon granules. The electrical resistance of the carbon changes with the different pressures caused by the vibrating diaphragm. The changes in resistance cause the current in the circuit to change. This changing current is amplified and transmitted to a receiver. At the other end of the circuit, a diaphragm in the receiver converts the varying electrical current back into sound.

For Ray Palmer to travel as explained in the pages of *Showcase*, he'd need to do more than shrink to microscopic size and jump into a phone receiver. The phone lines only carry electric current. Thus, Ray would have to convert his body into electricity to travel the wires. At the least, Ray would have to shrink his body to the size of an electron. To achieve that size, the atoms in Ray's body would have to be crushed so tightly together (see our Green Lantern chapter) that Ray would form a black hole. In other words, it's impossible.

Shrinking to the size of electrons raises another comic book concept that's still used in comics, although scientifically it's been outdated for nearly a century. It's the notion that the structure of atoms resembles that of a solar system. In *The Atom* #5, March 1965, in a story called "The Diamond of Deadly Dooms," a scientist explorer friend of Ray Palmer calls Ray and tells him to come by to see his latest exciting discovery. At the mansion, Ray witnesses a mysterious ray beam from inside a large diamond that turns his friend, and later a cat, into diamond structures. Ray turns on his shrinking power to full blast and shrinks to microscopic size. In one panel, he repeats an old theory, proven wrong approximately seventy-five years ago, that "There is a theory that some atoms may be entire solar systems on a submicroscopic scale."[38] Ray discovers an entire subatomic world inside a diamond. The people are the last survivors of the island of Atlantis, and on the subatomic world they live in peace and plenty, never dying. Needless to say, a few crooks and troublemakers in the real world and in the atom-sized world cause trouble, but the Atom puts a stop to their evil schemes. It's not just the Atom stories that feature whole worlds hidden in a lump of sugar or a galaxy in a teacup. Stories of transformation in size from the very small to the very large have always fascinated people. Thus mythology is filled with the tales of giant gods and insignificant humans, and of humans becoming as powerful as the gods trying to conquer them—elements that when used correctly make a fascinating story. Just not a scientifically accurate story.

A solar system has a sun in the center of the system with a number of planets rotating around it, each in a well-defined, stable orbit. Knowing the orbits of the planets and the speed at which the planets travel around the sun, we can locate every planet of the solar system at its precise position at a specific time and date.

[38]"The Diamond of Deadly Dooms," *The Atom*, March 1965.

An atom has a nucleus in its center. The nucleus is made up of neutrons, particles with no charge, and protons, particles with a positive charge. Circling around this nucleus are electrons, particles with a negative charge. Atoms normally have the same number of electrons and protons, so that their charges balance out. If atoms have more protons than electrons, or vice-versa, they are called ions.

Electrons speed around the nucleus very quickly. They move at a specific distance from the nucleus, but there can be more than one electron circling a nucleus at the same distance. The electrons don't smash into each other because they are repulsed by their electrical charges. However, the electrons can zoom around in many different orbits around the nucleus as long as they remain at the specific distance to the atom's core. These numerous orbital paths for electrons make up what is thought of as a shell around the nucleus. These shells aren't real, but just a specific distance from the atom's nucleus. Each of these imaginary shells around an atom's nucleus can have a certain number of electrons in it.

The first shell around a nucleus can have two electrons in it. The second shell around the nucleus can have as many as eight different electrons whizzing around in it, all at the same orbital distance but never colliding with each other. The third shell of an atom can have eighteen electrons in it, and so forth.

Already, we can see a major difference between a solar system and an atom system. In a solar system, one planet is all that can orbit the sun at a specific distance. In an atom system, the nucleus is surrounded by electron shells, with a specific number of electrons possible per shell. The electrons in each shell are all orbiting the nucleus at the same distance, but their speed and location are all different. In our solar system, we know the location of every planet at any time. That's definitely not the case with electrons in atoms.

The basic discoveries regarding atomic structure were made by Niels Bohr in 1913. It was Bohr who compared the structure of an atom to the structure of the solar system. While the comparison had some validity, it was only partially true. Bohr made the comparison so that people could have some understanding of atomic structure. It was never meant to be the total or final description of all atoms.

The Heisenberg uncertainty principle, introduced by the German scientist Werner Heisenberg in 1927, said that when describing an electron in an atom, it was impossible to specify *simultaneously* its position and its momentum with precision. The theory also stated that a more accurate determination of one of these facts would result in a less accurate description of the other. One of the foundations of quantum theory, Heisenberg's uncertainty principle made it clear that we could not accurately predict at any one instant where an electron was in relationship to the nucleus of the atom. If we determined where the electron was, then we wouldn't know how fast it was traveling.

Quantum mechanics demonstrated that electrons weren't at all like planets. Nor was a nucleus of an atom anything like the sun. Unfortunately, what was meant to be a simple example of a complex structure instead became accepted as the structure itself. Most people today, when asked about the structure of an atom, still believe that atoms resemble solar systems, with electrons as planets and the nucleus as the sun. Their knowledge of atomic structure dates back to the earliest Niels Bohr model.

If Ray Palmer could shrink to atomic size, he'd find himself adrift with a bunch of highly charged particles zipping around him—no place to live or start a kingdom, that's for sure.

Nor will someday a giant suddenly appear who is so big that he fills most of space. Not even if he continues to shrink until he is the size of a normal man, and then he shrinks himself into a subatomic world that's inside a dust mote he finds on the road.

Small is great fun. Small is interesting. Small makes for good stories and wonderful imagery. But, unfortunately, as we've learned in this chapter, small is impossible.

Chapter 8

# Fast, Fast, Fast

## The Flash

### Introducing the Flash

**The smash success** of Superman guaranteed that new superhero comics would follow immediately. A half dozen new comic book publishers opened their doors from 1939 through 1941. As always in publishing, imitation was the sincerest sign of success.

Not content with only Superman and Batman, DC Comics soon released several more magazines featuring the heroics of a whole new crew of incredible characters. Most had much less dramatic life stories or explanations for their powers than Batman or Superman, but all were straight arrows determined to use their unique powers in the battle of good against evil.

Unlike Superman, who was gifted with many powers, the Flash had only one power: he could move with super speed. The Flash's origin was explained in the first issue of *Flash Comics*, January 1940. Jay Garrick, a brilliant college biology major (but a mediocre football player) breaks a bunch of chemical beakers containing gas that emanates from "hard water." When pausing for a smoke, Jay accidentally knocks over the containers and is overcome by the fumes.

In the hospital, a doctor states that "science knows that hard water makes a person act much quicker than ordinary. Jay can walk, talk, run and think swifter than thought."

Jay uses his newfound powers to win the big football game for his college. Then, he's off chasing crooks, moving so fast that he's invisible to the human eye. In the same story, Jay also defeats a group of wartime profiteers by catching the bullets that they fire at him in midair.

The early Flash stories were mildly entertaining and mostly involved human crooks trying to outwit the Flash and his amazing speed. Stories were short and to the point and exhibited a reasonable amount of cleverness. Jay was a model citizen, but he rarely used his super speed other than to run ahead of the police and catch the criminals. Little attempt was made to highlight the more unusual gifts possessed by the fastest man alive. Still, super speed sufficed for readers, and the original Flash appeared in 104 issues of his own comic, as well as in *All-Flash*, *All-Star Comics*, and *Comics Cavalcade*. He even inspired several super-fast comic book imitators, including one from his own company called Johnny Quick.

Superhero sales slumped after World War II as readers embraced horror and crime comics. Funny animal comics, western comics, and war comics thrived as one after another the great superheroes of a decade earlier disappeared with hardly a notice. Comic sales were booming until the Senate held hearings to look into charges that linked comics with the rise in juvenile delinquency. Comic book sales slumped and it looked like the end of the comic book publishing field. A number of smaller publishers went under.

Reenter the Flash. In 1956, DC started a new comic, *Showcase*, featuring the adventures of a new character every month. A companion title, *The Brave and the Bold*, soon followed. The two titles served as a place to try out new and different heroes. If sales were good, the character got his own comic. If sales were poor, the character was abandoned.

The Flash was the first Golden Age comic character to appear in *Showcase*. The revived hero was edited by Julius Schwartz, who had edited the series for much of its initial run. This time around, Schwartz insisted the comic have a more believable scientific background to appeal to his young, more scientifically sophisticated readership. The first appearance of the new Flash in 1956 was the beginning of what comic book fans called the Silver Age of comics.

The new Flash was Barry Allen, a police scientist, who acquired his powers when a lightning bolt hit a rack of chemicals in his lab, dousing him with a mix of electrified solutions. The mix gave Barry super speed just like the original Flash (who just happened to be Barry's favorite old comic book character). In short order, Barry adapted to his new powers and even constructed a costume to wear when adopting the Flash persona. Since he was already a policeman, there was never a question that he would fight for law and order.

The new edition of the Flash was a huge hit, and, after several tryout issues of *Showcase*, the series was awarded its own monthly comic in early 1959. The Flash lasted in his own series for nearly thirty years before Barry Allen died a heroic death. However, not even death could end the reign of a comic book superstar, and other characters from the series took over as the Flash, as the series continued into the 1990s. The Flash even appeared for several years as the hero of a weekly television show.

Without question, the Flash is one of the most successful comic book characters ever created. But, despite efforts to ground his powers in scientific possibility, the Flash is yet another example of an entertaining but totally impossible hero. Though the Flash only has one super power—incredible speed—it's more than enough to ensure he'll never be anything more than an imaginary character. Let's see exactly why.

## Problems with Logic

It doesn't take a scientist to identify many of the problems involving super speed. Let's assume that the lightning bolt that hit Barry Allen and drenched him with chemicals gave him the power to move at incredible speeds. If Barry Allen could move faster than anything on Earth, would he be able to control his power and use this speed to become the Flash, the world's fastest crime fighter? Or would Barry discover his gift was no gift at all and actually a curse in disguise?

First, as a basis for comparison, let's look at running and speed in the real world. How fast can an ordinary human run? How fast *is* fast in regard to other animals and objects?

Running speed is directly related to how many steps you take (cadence) and how big the steps are (stride length). Or put into a formula format: Velocity = cadence × stride length.

Stride length is directly dependent on the length of the leg of the runner (length). Cadence is somewhat more difficult to calculate, as it depends on the acceleration of gravity, independent of speed, as well as the movement of the leg. Plugging in all the numbers and calculating, we come up with an estimate for the maximum running speed for a man: $V = \Pi (g \times \text{length} / 6)^{1/2}$—where g = 9.8 meters per second/per second.

Note that the only variable in the equation is the length of the runner's leg, leading us to conclude that tall people can run faster than short people. Distance is related to strength, stamina, and mass.

If we estimate that a tall runner's leg is a meter long, and we plug in all the numbers, we discover that the maximum running speed for all but the very fastest runners is approximately 4 meters/second, or just about 10 miles per hour.

Now, man is by no means the fastest runner in the animal kingdom. The fastest land mammal[39] is the cheetah, which can achieve speeds up to sixty-four miles an hour. We should note that cheetahs can only maintain this speed for approximately a hundred and fifty meters before fatigue sets in. One of the most amazing things about the cheetah is that it can go from rest position to a speed of nearly a hundred feet per second (forty-five miles an hour) in 2.5 seconds.

The cheetah's great speed is due to its unique body construction, with long legs and shoulder blades that are not connected to its collarbone. This enables the cheetah to increase both its stride length and its cadence in an instant. When running, a cheetah can cover twenty-three feet per stride and move at four strides per second.

---

[39]A few predatory birds dive more quickly than the cheetah can run. This is why we limit our discussion to *land* mammals. Certain falcons have been recorded diving at more than 125 miles per hour.

Three other adaptations help a cheetah run. They have very large lungs that rapidly move oxygen into their blood. They also have a large heart, which moves the oxygenated blood quickly from their lungs to their muscles. Cheetahs are different from all other members of the cat family in that their claws are not fully retractable. Thus, their claws are exposed while running. The cheetah's claws are extremely important for traction, enabling the animal to make quick turns when hunting.

Before turning our attention to Barry Allen, let's look at one more popular measure of speed: the famous Superman slogan from the 1940s, "faster than a speeding bullet." The speed of a missile or bullet is dependent on a number of factors. Gravity acts on a projectile, pulling it downward. The size, shape, and air density affects the speed of the bullet. Air resistance also slows down a bullet.

Still, a cartridge fired from a .22 Swift rifle travels over 1,200 meters/second, or three-quarters of a mile per second. That's a lot faster than most commercial airplanes that fly around 600–650 miles per hour, or a mile every six seconds.

Armed with these facts, let's examine the speed of our hero, the Silver Age Flash Barry Allen, and see how he measures up.

In "The Mystery of the Human Thunderbolt," the first adventure of the Silver Age Flash, Barry Allen sees a bullet fired at his girlfriend, Iris West, and he pushes her out of its path just before it strikes. This means that Barry's vision and reaction time are faster than a bullet's speed, or 1,200 meters per second. Later, in the same story, when he chases the master criminal known as the Turtle, Barry breaks the sound barrier, a mere 331.29 meters per second.[40] We should mention that the story also displays Barry running at "eye-blurring speed" at a wall, then coming to a stop without effort. In the climax of the story, Barry runs on water, explaining to himself (and the reader) that he's moving so fast that his feet don't have time to sink into the water.

---

[40]His breaking the sound barrier is picked up on radar, which leads us to wonder why the government had a radar screen focused on the ground.

Within a few months, Barry is running across the Atlantic Ocean (*The Flash* #107, June 1959, "Return of the Super-Gorilla"), moving so fast that he doesn't have time to sink. Even more impressive, in the same story, he spins around so fast that he bores into the earth as "the particles of his body slip between the atoms of the solid earth."

In "The Conquerors of Time" (*The Flash* #125, December 1961), the Flash and Kid Flash (Wally West, a teenager with the same super speed as the Flash) travel forward and backward in time using a "cosmic ray powered treadmill." The Flash explains that the machine is coordinated with radiation pulses from a cosmic ray clock. When the treadmill goes forward, positive cosmic ray radiation is released, sending the Flash into the future. When the treadmill goes backward, negative radiation is released, sending him backward in time.

Advertisements of the time mention how the Flash moves as fast as a lightning bolt (the speed of light, 300,000 km/sec). Then, of course, there's the adventure (actually several adventures) in which Wally West travels *faster* than the speed of light.

Assuming the Flash is the fastest being in the universe, could he do everything described? Or are there scientific problems involving great speed that his creators conveniently ignore?

The most basic problem, of course, with all Flash stories is the astonishing speed at which he moves. If the comic book world was real, none of Flash's adventures would ever happen. That's because the Flash is so fast, he moves with such incredible speed, that there's no possible way that the villains could ever pose a challenge. By the time they realize he's after them, they'd already be in jail. After all, even when he's just cruising along, our hero's running nearly at the speed of sound.[41] This happens to be nearly a thousand feet per second. When we're paying close attention to something, our eyes blink on the average once every four seconds. Many scientists believe that blinking is the moment it takes for our brain to

[41]He can run much faster, but we'll assume that most of the time he tries to keep his speed down to avoid sonic booms.

assimilate information. But in the case of the Flash, our eyes act much too slowly.

In those few seconds, the Flash has moved nearly a mile from where we saw him. Human reaction time under extreme stress ranges from about 0.1 to 0.3 seconds. If our villain reacts to the Flash at even the fastest possible speed, firing his weapon at the Flash, our hero will already be a hundred feet away—or he will already have bound, gagged, and thrown the crook over his shoulder.

The truth about the Flash is that even if he is only a fraction as fast as claimed by his writers, he's never going to be really challenged by any criminal scheme. He is capable of seemingly magical solutions to his worst disasters by just turning up the speed a notch or two. In "The Mirror Master's Magic Bullet" (*The Flash*, #119, March 1961), the evil Mirror Master fires his gun at a mirror the moment the Flash opens the door to his secret hideaway. The vibrations of the mirror breaking when hit by the villain's "magic bullet" paralyze the Flash. Why would the Flash let that happen? The Mirror Master makes it clear that he can't fire his gun until the Flash enters the room (due to some bizarre technobabble mirror science). But letting the Flash enter a room is letting a genie out of a bottle.

As mentioned earlier in this chapter, a bullet travels at 1,200 meters per second. The Mirror Master is perhaps a meter away from the mirror reflecting the Flash's image. The Flash looks to be ten meters or less from the same mirror. It takes the bullet 0.0008 seconds to hit the mirror. So, if the Flash runs a meter in 0.00007 seconds, or ten meters in 0.0007 seconds, he'd be able to catch the troublesome bullet without breaking into a sweat. Doing some quick calculations, we discover that this translates into approximately 14,285 meters per second, or 14.28 kilometers per second. That sounds incredibly fast until we remember that the speed of light is 300,000 kilometers per second. Of course, we wouldn't be playing fair if we didn't remind everyone that the escape velocity on Earth is 11.2 km/second. So, if the Flash were running up a long ramp instead of inside a small cabin, he'd launch himself into outer space.

As demonstrated above, super speed guarantees that there's no challenge that the Flash can't handle—except for those affecting his own body.

The writers of *The Flash* tried from time to time in the series to explain away the Flash's problems dealing with the real world. For example, they claimed that friction didn't affect the Flash due to a special high-energy shield that formed around his body while running. This was a good idea, because running at such tremendous speeds would quickly wear out the best sneakers, not to mention his costume.

Still, no friction would also be a problem. Friction helps people convert one form of motion into another. When the Flash runs, friction is what converts the energy of his feet pushing backward against the ground into forward motion. The only explanation we have for the Flash running on water is that his feet move so fast that they generate friction against the surface tension of the fluid, and push him forward like a surfboard.

What never seems to be mentioned by the writers of *The Flash* is how much he eats. The lightning potion gave Flash incredible speed, but it didn't provide him with an infinite energy supply. That's impossible by the three laws of thermodynamics. All that constant running at high speed must burn some serious calories.

The calories burned in running (or any exercise) are a function of body weight, the intensity of the workout, conditioning, and metabolism. It's estimated that a man weighing 160 pounds and running ten miles in an hour burns 1,267 calories. But ten miles isn't much of a distance when compared to the Flash running from Central City to Africa, somewhat more than three thousand miles. The number of calories burned increases with speed and distance. It's not inconceivable that Flash would burn 375,000 calories or more on his trip to the jungle. Normally, an active man requires about 3,000 calories per day, meaning that Flash must have eaten some meal before leaving.

On a related note, does the Flash sweat? If not, why not? If so, then his uniform must be fairly porous. All that running might steam off the sweat, but the chemical residue might make the Flash the superhero whose body odor announces his presence before he arrives!

Another problem that's related to the Flash's speed and perceptions is our hero's ability to hear. Since he is often running faster than the speed of sound, how does he hear anything? In stories like "The Mirror Master's Magic Bullet" and "Trail of the False Green Lanterns," he's obviously traveling faster than the sound waves carrying messages to him from other characters. For that matter, why isn't Central City's City Council up in arms about the Flash constantly causing sonic booms? How would you like living in a city where there's a boom or two every day?

What about vision? How does the Flash manage to see while running at speeds well over a thousand miles a minute? How can he even keep his eyes open in winds that are many times worse than the most terrible tornadoes? For that matter, wouldn't that same wind ravage his face and skin? Even a small dust cloud would be dangerous moving at such enormous speeds.

Much worse in the vision situation is how the Flash can react quickly enough to deal with dangerous situations. When he's running down the side of a building (defying gravity by speed, according to the editors), he's moving at an astonishing speed. The ground's rushing closer at many thousands of feet per second. How does he know when to turn so that he doesn't smash himself flat on the sidewalk?

For the Flash to function at such enormous speed, his body needs to act at those same speeds. He needs to breathe at super speed, see at super speed, and, as made clear by our last paragraph, react at super speed. There's only one problem. There's a limit to how fast humans can react to events. The neurons in our mind can't operate faster the speed of electricity, the speed of light. Though the Flash can move thousands of times faster than an ordinary human, even he needs a few microseconds to observe and comprehend a problem. And at a speed of many thousands of meters per second, he doesn't always have the time.

Another basic problem never discussed in *The Flash* adventures is momentum. The Flash has the fastest first step in history. He accelerates faster than any object in the universe. Our hero is rarely shown building up speed. Instead, he starts running, and after one or two steps, he's moving at fantastic speed. What happens, though, when he comes to an abrupt, unplanned stop—like when he crashes into a wall in "The Mystery of the Human Thunderbolt" or into a mirror in "The Mirror Master's Magic Bullet"?

And just how much power is in his punch?

Linear momentum is the fundamental quantity defining the motion of an object. It's equal to the product of the mass of an object multiplied by the (linear) velocity of the object. In an isolated system with no outside forces, such as the Flash running into a wall, total momentum remains unchanged over time. This property, one of the basic laws of physics, is known as the conservation of momentum. It's valid not only in our everyday world but also in quantum mechanics (dealing with the very small) and relativistic mechanics (dealing with systems moving near the speed of light). Newton's second law of motion says that a force acting for a certain time produces a change in momentum.

It's time to plug in some numbers and see how much energy is released when the Flash hits that wall, using this equation: Momentum = mass × velocity (or $p = m \times v$).

To keep things simple, we'll have the Flash weigh 75 kg, about 165 pounds, and we'll say his velocity is 1,500 meters/second (a little faster than the speed of a bullet). Plugging the numbers into the formula, his momentum comes out as 112,500 kg-meters/second.

Newton's second law says that the force (F) experienced by the Flash over the time (t) it takes him to stop is equal to the mass of the Flash times the velocity. In other words: $Ft = mv = p$ (or $F = mv/t$).

If the Flash collides with a wall in one thousandth of a second, we arrive at a force of 112,500,000 kg-meters/second$^2$. A newton is defined as the force that produces an acceleration of one meter/second$^2$ on one kilogram, so our force is equal to 112,500,000 newtons. In simpler terms, that's enough force to start Air Force One rolling down the runway. That's a lot of force.

## The Speed Barrier

Most of the problems we discussed in the previous section were ones of simple logic. Most obvious questions are difficult to answer in the pages of a comic book, where the story is devoted to action. Most questions merely take normal everyday situations and raise them to another level—how does the Flash consume enough calories to keep his body moving, how can he navigate when running so fast, and so forth. These are problems brought about by a unique scenario, and as such, they might be solved by unique solutions. That's one of the basic premises of superhero comics—coming up with real-world remedies for imaginary troubles.

However, there are some problems that aren't so easy to dismiss. A few problems arise from matters that can't be dismissed with a dollop of super science and a magic crayon. Maybe an imaginative writer or editor can figure out how the Flash can get enough food to keep running at high speeds for hours. The answer might be improbable, but it is still an answer. Unfortunately, some problems don't have answers, no matter how inventive the writer or editor might be. The laws of physics, for example, are laws. And as such, they cannot be broken.

According to Einstein's theory of relativity, nothing in the universe can exceed the speed of light. That speed, measured in a vacuum, is approximately 300,000 km/sec (actually, 299,792,458 meters/sec, to be precise). Light is part of the electromagnetic spectrum, which includes infrared radiation, ultraviolet radiation, gamma rays, X-rays, radio waves, etc. All of these travel at the speed of light.

Equally important, the theory of relativity states that light travels at the same speed relative to an observer regardless of the motion of the observer. Thus, if a man was traveling in a spaceship at half the speed of light, light from the spaceship would still travel at the speed of light.

Nothing can go faster than the speed of light. That's a rule that can't be broken in life, although it is often broken in comic book stories. If the sun suddenly exploded, we wouldn't see it happening. The light from the explosion would take eight minutes to reach us. Which is exactly the same time it would take the heat and radiation to reach us and fry us to ashes.

The speed of light isn't some sort of wall that exists that can be climbed over or tunneled through. Good science fiction makes this fact clear. Starships in *Star Trek* or *Star Wars* never travel faster than the speed of light. Instead, these ships move through warps in space that enable them to travel from one distant point to another by taking a shortcut through the fourth dimension. The ships have "jumped" from one location to another. They haven't flown the entire distance.

While space warps sound confusing, they're actually quite easy to understand through example. Take an ordinary piece of white paper and draw a circle on the bottom of the page and a star at the top of the page. The circle is our spaceship on Earth and the star is our destination. We want to travel from Earth to the star. According to mathematics, the shortest distance from one point to another is a straight line. So, under normal circumstances, our trip would cover ten inches of paper. If it took a minute to draw an inch (we are holding the pencil with our toes), then it would take us ten minutes to travel from the Earth to the star.

But what if we could warp space—or as it is often called, "fold" spacetime? What if on our spaceship we had a device—call it a warp drive—that could fold the paper in half, so that the circle touched the star? Then our distance would be only a few thousandths of an inch, taking nearly no time at all. Our ship didn't go any faster, nor did we travel ten inches along the paper. Instead, we merely used our warp drive to take us from one location to another.

What we did was take two-dimensional space and fold it in three dimensions. If we somehow can find a way to fold three-dimensional space in a fourth dimension, then we could travel throughout the galaxy without ever flying faster than the speed of light.

Still, you might ask, why can't we just fly our starship very quickly and move between our two points? What if we don't know how to bend space (which at present we don't)? What then?

Then, our fellow Earthlings, we don't travel to the stars. Or if we do, we prepare for a very long trip, because the nearest star is four light years away. It would take us four years to reach our destination if we could somehow get our spaceships to move at the speed of light.

Leading us to finally to throw up our hands in annoyance and ask "What makes the speed of light so special?" Why can't anything move faster than light?

The answer comes from Einstein's Theory of Special Relativity. Please remember, as mentioned in our introduction ("A Word about the Law"), that this theory has been proven time and again by scientists studying everything from the smallest objects in the universe (electrons) to the largest (galaxies of stars).

Einstein defined the relationship between mass and energy in his famous equation: $E = mc^2$, where E = energy, m = mass, and c = speed of light. Mass measures the amount of inertia an object has, that is, the resistance the object offers to a change of its state of motion. Therefore, when an object starts moving, increasing its kinetic energy, the mass is directly affected. The greater the speed and thus the energy, the greater the mass. However, since we're working with 300,000 km/sec squared, changes in mass are going to require huge changes in speed.

Doing some calculating, we come up with the following formula to calculate the mass of an object in motion: $M_m = M_r / \sqrt{(1-v^2/c^2)}$, where $M_m$ is the mass of the object in motion, $M_r$ is the mass of the object when it isn't moving (at rest), v is the velocity of the object, and c is the speed of light. This equation is a direct result of Einstein's special theory of relativity.

Let's look at the conclusions we can draw from the equation.

When an object isn't moving, v = 0, and the mass of the object remains unchanged. Since c = 300,000 km/sec (approximately 186,000 miles/sec), v needs to be huge before the amount $v^2/c^2$ has much meaning. For example, let's assume the Flash has the velocity slightly above that of a speeding bullet, 1,500 meters/second, as described earlier. That's about 5,400 km/hour (3,300 miles/hour). Plugging the numbers into our equation, v = 1.5, so v/c = 1/200,000,

$v^2/c^2$ is $(1 / 4 \times 10^{10})$, and that number is so small that the equation is pretty much $M_m = M_r$.

In other words, mass doesn't really change very much until the moving object approaches the speed of light. When we say approaches, we mean *significantly* approaches. Again, resorting to example, let's take the case where the Flash is moving at one quarter of the speed of light, 75,000 km/sec—fast enough to run around the Earth 112 times a minute! Doing our calculations, we find that the mass of the Flash increases by 16/15 his original mass. If the Flash weighs 165 pounds, as postulated earlier, he'd thus weigh 176 pounds. Not a major change for traveling so fast, and surely not enough weight to slow him down much.

Clearly, as our velocity gets closer to the speed of light, $v^2/c^2$ gets closer to 1. This means that $1 - v^2/c^2$ gets closer to zero. The square root of a number approaching zero also approaches zero. As our denominator approaches zero, mass approaches infinity.

In simple terms, if the Flash could run at the speed of light, he'd have infinite mass. While that might sound interesting, it's impossible, because that would mean he'd have all the mass of the universe. If he can't run at the speed of light, he can't run faster than the speed of light. Not in our universe.

So forget the speed force in the Flash stories. Forget about that scene in the Superman movie where he flies faster than the speed of light to travel backward in time. It's not going to happen. Nobody can move *at* the speed of light. Nobody can move *faster than* the speed of light.

## Chapter 9

# Good, Evil, and Indifferent Mutants

## The X-Men

### A Victory Snatched from the Ashes

**In 1963, Marvel** was turning out comic book after comic book, with most of them succeeding well beyond anyone's expectations. Stan Lee, especially when working with Jack Kirby, seemed to have the magic touch. So, when a new team comic book titled *X-Men* appeared, readers and fans expected it to fly as high as other projects from the same period like *The Avengers* and *Daredevil, the Man Without Fear*. Yet despite the high hopes, *X-Men* didn't seem to capture the same magic as many of the other Marvel titles.

The comic was entertaining, with a premise unlike that of any of the other series Marvel was publishing. Yet large parts of it seemed too similar to other Marvel comics already being issued. Not helping the series was that Jack Kirby and Stan Lee didn't work on the comic very long; instead they passed along the script and art duties to other Marvel employees. Early writers on the series included Roy Thomas and Gary Friedrich, while artists included Werner Roth and George Tuska. The X-Men had their own comic, but they didn't have a genius writer or genius artist to propel them from the middle of Marvel's pack to the front lines with *Spider-Man* and *The Fantastic Four*. Yet comic history often delivers when least expected.

The basic premise of *X-Men* was similar to a number of science fiction novels that had appeared during the aftermath of World War II and the atomic bomb. Actually, a recent X-Men limited series used the same name as the most popular novel about mutant children written during that period, Wilmar Shiras's *Children of the Atom*. The novel (as well as the comic book series) dealt with mutant teenagers whose powers develop at puberty; they then find themselves feared and hated by normal people. The only way they can develop into healthy, happy human beings is to be raised at a special school or academy where they are taught in secret how to control their powers and how to blend in with the mundane humans who surround them.

The headmaster of this special, hidden school in *X-Men* was Professor Charles Xavier, or Professor X, wheelchair-bound but the most powerful telepath on Earth. Xavier called his recruits the "X-Men" because they derived their mutant powers from the X gene in their bodies. Members of the original X-Men team included Archangel, Iceman, Marvel Girl, Cyclops, and the Beast. Since no series about heroes was complete without major villains, the X-Men fought against others of their own kind who were out to subjugate mankind and establish an empire ruled by mutants. This Brotherhood of Evil Mutants was run by Magneto, the Master of Magnetism. In early adventures, most of the fights between good and evil mutants took place in secret, but after awhile, they spilled into the open, raising the fear factor among ordinary people that the side winning the mutant battle would someday control the world. It was the beginning of Marvel's longest running plot line.

In the Earth of the Marvel universe, normal people feared and hated mutants with at least as much intensity as that shown by reality's most bigoted racists and extremist groups. A move to a race war between man and mutant was built up month by month in the *X-Men* comics for nearly forty years. In 2001, it finally appeared ready to happen.

As described by the original series, Professor X was the world's greatest telepath. His mental powers were so strong that he could take control of a person. He could make the person obey his commands even though that person was nowhere near him. Every member of the X-Men team had a code name that matched his or her super power. Thus, Archangel, Warren Worthington III, had wings and could fly. Cyclops, Scott Summers, shot deadly power beams from his eyes. Jean Gray, Marvel Girl, was a telekinetic and also a telepath. Iceman, Bobby Drake, could cover anything with ice, including himself. And Hank McCoy, the Beast, was a scientific genius with the power, stamina, and reflexes of a wild animal.

Although they battled the usual run of Marvel villains (and sometimes heroes, as editor Stan Lee knew the value of plots that had one group of heroes fighting another through some sort of mistake or misunderstanding), the X-Men primarily fought other mutants. Villains included the Blob, the Mimic, Banshee, Polaris, and numerous other characters, many of whom later changed direction from evil to good and became members of the X-Men lineup. Of all the villains to appear in the early series, perhaps the most daunting were a group of giant mutant-hunting robots called the Sentinels. Nearly forty years after their initial appearance, the Sentinels still hunt the X-Men in the pages of *Ultimate X-Men* and *New X-Men*.

Still, the earliest version of *X-Men* never seemed to catch fire and in March 1970, new stories stopped appearing. Instead of canceling the comic, the powers that be at Marvel continued *X-Men*, reprinting stories from earlier issues. This policy continued until May 1975, when the company published *Giant-Size X-Men* #1, featuring the all-new, all-different X-Men.

The new team, created primarily by writer Len Wein and artist Dave Cockrum, featured Professor X and Cyclops from the original team. Working along with them were five new X-Men. Storm was a regal black woman with the power to control the weather. Colossus was a Russian who could change his body to living steel. Nightcrawler was a blue-skinned circus performer who could teleport. Thunderbird was a Native American with aerial powers. And Wolverine was a Canadian secret agent with incredible healing powers and retractable metal claws.

The new team took over from the original X-Men with issue *X-Men* #94, August 1975. That issue saw the resignation of most of the original team members, as well as the death of Thunderbird. More important, it began the run of new writer Chris Claremont.

Over the next few years, Claremont, working first with Dave Cockrum and later with John Byrne, established *X-Men* as the most exciting, innovative comic book being published. The book, with its continuing plot line and complex characters, slowly but surely attracted a larger audience. Claremont's Dark Phoenix saga drew rave reviews, and his follow-up story, "Days of Future Past," established him as the most popular writer in the comic book industry. In an incredible run of creativity, Claremont continued writing X-Men comics into the early 1990s, for a record eighteen-year span.

With astonishing regularity, *X-Men* has stayed at the top of the charts of comic book sales for the past twenty years. Though the comic has gone through a number of writers, the stories continue to fascinate fans, who have made it the best-selling superhero comic book in the world.

Mutants are real. They exist and have existed since the appearance of life on Earth. Mankind has evolved and changed over the ages. We're still evolving. At its core, that's what *X-Men* is all about: how much we change and yet how much we remain the same.

Over a period of nearly forty years, *X-Men* and its many spin-off titles, including *X-Force*, *Uncanny X-Men*, *Astonishing X-Men*, *Generation X*, and *X-Factor*, showcased hundreds of mutants with incredible powers. Every month, some new super-charged human came face-to-face with Xavier's students, forcing them to fight the good fight and win the good war. While it was stated repeatedly throughout the comics that mutants formed only a small percentage of the world's population, they all seemed anxious to relocate in New York City or its surrounding area. More than once, mutants made up the entire cast of an issue of *X-Men*, with not a solitary ordinary human to be found. Although they were a small percentage of the globe's population, mutants were everywhere.

With rare exceptions, every mutant in the X-titles had his or her powers spelled out in great detail. Oftentimes, the character's special ability made little sense other than that it helped satisfy some element of the storyline. In answer to the standard question of which came first, most often it was the plot before the mutant and not the other way around. Mutants were created by writers for the story being told. They were beings of convenience, not logic.

Other authors have written at length about individual X-Men and how their powers might work. Most theories center on what little knowledge we possess about parapsychology and the human brain. Telepathy, perhaps considered to be the most common ESP (extrasensory perception) talent, has been studied by scientists for nearly fifty years without a major breakthrough. If researchers haven't been able to explain how people communicate telepathically, how can anyone devise a rational theory about how Magneto exercises absolute control over magnetism?

Instead of focusing our attention on questions that can't be answered, we want to look at the bigger picture, the main concept that underlies *X-Men* and every other comic that's been written about superhuman mutants in our midst. It's a question that's become increasingly important as the barrier between politics, religion, and science continues to crumble. What exactly is evolution? Is modern man the product of evolution? If it's true, then why is it still called a theory? If evolution's true, is man still evolving? Are mutants the next step in evolution's ladder? As asked by one of the characters in *X-Men*, the movie, will homo superior someday replace homo sapiens? Are the X-Men our future—perhaps even our children?

## The Case for Evolution

Evolution in this chapter refers to a process that involves living things, not objects. Talking about how animals evolve is obviously much different from discussing the evolution of suns, as discussed in our chapter about the Green Lantern. As far as we're concerned in this X-Men discussion, evolution is a biological change in a group of living creatures over a period of time greater than the lifetime of any individual being.

Evolution is usually broken down into two processes, micro-evolution and macro-evolution. Micro-evolution refers to the evolutionary changes within a species. Macro-evolution is the evolution of one species into another. For example, micro-evolution regarding whales would be their gradual change from freshwater mammals to saltwater mammals. The macro-evolution of whales would be land animals evolving over millions of years into whales, or vice versa.

What exactly is a species? In the most basic terms, it's a group of creatures that breed among themselves. When a species is wiped out, it's called extinction. When a new species comes into existence, it's called speciation. Humanity is a species. Most of the issues of *X-Men* deal with micro-evolution rather than macro-evolution.

Charles Darwin first proposed his theory of evolution in 1859. In his theory, Darwin noted that there were biological variations among individuals of a species. Darwin didn't know what caused these variations, but since that time we've learned that such variations were caused by mutations, or physical changes in a gene or chromosome. A few variations were useful while many were not. Darwin further noted that because of limited resources and continued population growth of a species, there would be competition between members of the species for resources. This competition would lead to the death of some members. Darwin concluded that members of the species with *useful* variations *would be more likely to survive* and reproduce than those without the useful variations. He called this concept *natural selection*. Darwin thus concluded that evolution was a result of mutations and natural selection.

Darwin's theory revolved around his observation that there were variations within a species, the process known as mutation. The DNA of any specific organism contains all the information needed to create the different cells necessary for life. So, simplifying explanations, mutations are the result of changes in the DNA code from generation to generation of a particular species.

Identical twins have identical DNA. So do clones, as we discovered back in our chapter about Spider-Man. Fraternal twins have less similar DNA than identical twins. Non–family members have DNA less in common than family members do. More important to our discussion, children inherit a mix of DNA from both of their parents. As we spread our net wider, we soon discover that every species has DNA that is different from the DNA of another species. Yet, even the DNA of different species have points in common, a subject we'll return to later in this chapter.

Because of the complexity of a cell, many types of mutations can occur. They can be harmful, neutral, or helpful. Most mutations are neutral. Harmful or helpful mutations often depend on the environment in which an organism lives. This point is demonstrated numerous times in the adventures of the X-Men, where team members use their mutant powers to help them survive dangerous situations where normal humans would be killed instantly. Although such gifts as Cyclops's eye blasts would be a harmful mutation if he lived in a primitive society, his use of it to blast robotic Sentinels intent on his destruction makes it a helpful mutation in our modern world. Taken as a group, the X-Men and their enemies use helpful mutations. Rarely in the series are neutral mutations seen (neutral presumably meaning boring to the writers). A few characters with harmful mutations appear in stories, but they rarely survive their initial appearance.

Darwin's theory of evolution caused a major stir when it was introduced in England in 1859. Opponents of the theory included many religious groups. At the time, a large number of people throughout the English-speaking world believed the word of the noted church theologian Bishop Ussher, who had determined through rigorous study of the Bible that the exact date of the creation of the Earth was October 22, 4004 B.C.E. Thus by Ussher's calculations our world was approximately 6,000 years old. Needless to say, Darwin's theory was in direct conflict with the bishop's date, as Darwin speculated that most species of animals living on the Earth, including mankind, had taken hundreds of thousands, if not millions, of years to evolve into their present state.

Within a short period, a war of words between those who believed Darwin and those who didn't began. Fortunately, the conflict rarely went beyond the talking stages. Even the famous Scopes trial, held in Dayton, Tennessee, in 1925, in which a teacher was prosecuted for teaching evolution in the classroom, generated plenty of talk and news but changed few opinions. Most biology textbooks in the United States barely mentioned evolution until the 1960s.

Sputnik and the Space Race changed all that. In the early 1960s, it was widely believed by the government and most educators that the Russians were beating us in the space race because of better schooling in science and technology. A long, hard look was taken at textbooks, and biology instruction changed to firmly embrace evolution. Opposition to evolution remained scattered and quiet.

However, strict believers in the words of the Scriptures refused to melt away. As the government continued to bar religion in classrooms, Evangelical Christians switched tactics. For years, they had objected to teaching evolution because they considered it a scientific religion, a part of secular humanism that stood directly opposed to organized faith. Instead of attacking science, the Fundamentalists became part of it. A number of theologians and scientists started promoting a concept they called *creation science*, in which they challenged the basic concepts of the theory of evolution as being unproven and unprovable. This new science declared that the theory of evolution was just one of many possible ways of interpreting events in the world of nature, and that many of the facts used by evolutionists weren't facts but opinion. The "creation scientists" offered their own version of how life began on Earth, using the Old Testament as their only source book. What at first seemed a desperate, last-ditch effort to force schools to acknowledge the story of creation in science classes soon developed into an all-out battle. In the ensuing debate, Fundamentalists argued that the evidence (the "science") that they presented proved that evolution was false. They asked that teachers be allowed to present Darwin's theory of evolution only if it was clearly labeled a "theory" and nothing more. More important, they demanded that if evolution is taught in a school, their alternative viewpoint of "creationism" be taught as well. Creationism, which was strongly rejected throughout most of the Western world, was embraced by a large section of the American population.

The battle between evolution and creationism came to a boil in the late 1990s, when the members of the state school board of Kansas voted to rewrite the school board's science education policy. Students would no longer be tested on their knowledge of evolution. Since teachers rarely taught subjects that weren't tested, and students only studied the subjects being tested, it meant that evolution dropped out of sight in the state's schools.

The dropping of evolution in Kansas schools led to a storm of protest across the country. College professors declared that students graduating from Kansas schools wouldn't be prepared for college courses because their knowledge of science would be seriously flawed. Scientific organizations refused to meet in the state. The state school board became a popular topic for late-night TV comedians. People throughout the state complained that Kansas had become the laughingstock of the country.

Meanwhile, Evangelical Christians, who had pushed the change in the school board rules, accused everyone from the major newspapers to the TV and radio stations of attacking their beliefs. Creationists trotted out their dog-and-pony show, claiming that the liberal politicians and big-media players were out to quash a viewpoint different from their own, even though the points raised by creationists, they said, were obviously true to anyone with common sense.

The voting public made their view known. Within a year, school board elections pushed three of the four conservative Christians off the school board, and the Kansas evolution ban was rescinded. But none of the old arguments were settled, and neither side backed down from its stand on evolution.

## The Truth about Creationism

The X-Men stand or fall with Darwin's theory. If God created everything at one time (4004 B.C.E) and God created man in his own image, man is already as perfect as God intended. Mutations in humans would demonstrate that man wasn't perfect when he was created. Moreover, the notion of humanity evolving upward into an even greater species suggests that one species can evolve into another, a claim that creationists say is absolutely impossible. So, let's examine creationism and see if what it claims as truth is scientifically accurate.

The main thrust of creationism is an attack on the theory of evolution. If the theory's not true, the creationists declare, then any conclusions drawn from it are equally tainted. Proving evolution not true would be a cataclysmic event in the history of modern science and technology. Much of our understanding of the physical and biological world relies on concepts derived from the theory of evolution. Remove one card at the bottom of a house of cards, and the house collapses. Creationists have hammered at this conclusion for decades.To them, it's vital evidence in their fight against evolution. By their reasoning, it's the smoking gun, the powerful *motive* that explains why scientists are willing to go to any length, to falsify any data, to prove that evolution is true.

Unfortunately for the creationists, though they have found evidence of phony fossils and other deceitful activity by a small number of people aimed at making a quick buck, they've yet to find a single hard *fact* that proves Darwin wrong. Despite all of their trying, the protestors are forced to use hyperbole, oftentimes inaccurate and misleading information, and arguments based on "obvious" material to make their case. In regard to appeals based on obvious beliefs, it's been observed, "the more you know, the less becomes obvious." Creationists rely too much on *facts* they claim are obvious—but can't be proved.

As we discussed earlier in this book, the word *theory* generates much of the heat when discussing evolution. One of the standard arguments made by creationists is that Darwin's work *is merely a theory, not a law*, and thus should not be treated as fact. We've dealt with theories before, but this time let's go through the step-by-step process that scientists use to prove that a scientific hypothesis is true.

Deducing and proving a theory consists of several stages:

1. Observe an unexpected or unusual occurrence.

2. Gather as much evidence as possible about the event.
3. Come up with one or more reasons (hypotheses) that might explain the observation. Come up with these explanations through analytical methods, trial and error, and detailed examination of the facts.
4. Design a test that will give predictable results if the hypothesis is true.
5. Conduct the test and check the results. Determine if the results demonstrate that the hypothesis (the explanation) is true.
6. If the hypothesis is not true, start over with a new hypothesis to explain the gathered data and test it again using steps 4 and 5. To be considered a viable theory, the same experiment or procedure must provide the same result.
7. If successful, publish the results in a peer-reviewed magazine.
8. Others achieve the same results when they follow the same procedure as defined by the article, confirming that the conclusions are reproducible.

After going through all these stages, especially stage eight, in which other scientists duplicate the same test independently and arrive at the same conclusion, a theory has been discovered.

Creationists like to use double-talk about this type of scientific study. They claim that by looking at processes, then stating a theory, then making a prediction based on the theory and checking to see if it is true, you are merely using circular logic to prove the theory. Needless to say, they conveniently forget to mention that scientists study numerous processes before formulating a theory; that once a theory is suggested, they test unexamined processes to see if they act in the manner predicted by the theory. Many theories are proposed in science, but only a few are proven correct.

Theories are proven wrong when new and unexpected results are achieved from the original test. Along with theories that didn't work, there is an entire class of scientific theories that stand on shaky ground, where results appear similar but not exactly what was expected. Such theories need a lot more work before they can be accepted as truth. Many theories are little more than speculation based on the slightest amount of corroborative evidence.

Other theories, like the theory of general relativity, the laws governing electricity, and the three laws of thermodynamics, have been proven so many times that no one doubts them.

Evolution is another theory that has been proven time and time again. A vast majority of the world's scientists believe it is true, as do most religious leaders and theologians. In 1996, the Pope released a formal statement to the Pontifical Academy of Science stating that "fresh knowledge leads to recognition of the theory of evolution as more than just a hypothesis." Any other theory with such overwhelming support would be considered true in an instant. Only the theory of evolution, which according to Fundamentalists contradicts the unalterable truth of the Bible, has faced so many challenges for 140 years.

Creationists argue that "every single concept advanced by the Theory of Evolution—is imaginary."[42] Another well-known Fundamentalist minister offers $250,000 for "scientific evidence that will prove evolution."[43] The prize remains unclaimed, since the proof requires a judge with an open mind, not a panel convinced of the infallibility of the Bible. Creationists have learned that the easiest method of dealing with evidence against their point of view is to ignore it or label it worthless.

Creationists state that most people believe in evolution because they have been cowed into submission by a vast network of scientists, atheists, and communists working in concert to discredit religion. The Fundamentalists like to call evolution the "Big Lie," as if scientific facts are mere propaganda. However, creationists exhibit the exact same type of behavior when they issue blanket remarks implying that their view is slowly gaining validity among scientists.

"Scientists who utterly reject Evolution may be one of our fastest-growing controversial minorities,"[44] proclaim numerous

---

[42]Cohen, I. L., *Darwin Was Wrong*, New Research Publications, 1984, p. 209.

[43]"Evolution Reward," *Battle Cry*, Sept./Oct. 1999.

[44]Hatfield, Larry, "Educators Against Darwin," *Science Digest Special*, Winter 1979, p. 94.

Internet sites. Not mentioned anywhere on those sites is the Gallup poll from 1997 that shows that 95% of scientists in the United States believe that evolution is fact. Surveys taken by scientific organizations throughout the world place the number of scientists believing in evolution at 99%. In 1986, a brief was filed with the Supreme Court to overturn a Louisiana law requiring balanced treatment of evolution and creation in state schools. All seventy-two Nobel Prize winners living in the United States signed the brief.

The main difficulty facing creationists is that the centerpiece of their belief system is the certainty that everything stated in the Bible is God's word. Thus, they believe that the Earth was created by God in six days, some six to eight thousand years ago. By assigning a specific day and time to the creation of the world and the universe, the creationists often find themselves rewriting scientific discoveries made in other fields to fit their one-size-fits-all absolute and unchanging timeline.

Creation scientists not only believe that God created the world in 4004 B.C.E, but, by implication, that he created the entire universe at the same instant, locating distant galaxies millions of light years away from Earth so that they would correspond with modern relativity theory and astronomy. Millions of years of fossil remains are described as residue from one great flood taking place only several thousands of years ago, while carbon dating is dismissed as totally false and inaccurate. In other words, they accept only data that fit their conclusions and reject any information that contradicts their belief.

In the 140 years since the theory of evolution was proposed, no real challenge to its validity has been proven. The theory of evolution fits in with every other scientific discovery we've made about the universe. Consider this evidence.

Charles Darwin predicted that the ancestors of trilobites would be found in pre–Silurian Age rocks. His prediction turned out to be true, as they were later found.

In 1859, Darwin said that the total lack of Precambrian Age fossils was unexplainable and that the lack could be a strong argument against his theory. However, such fossils were discovered in 1953.

They had been around all along, just too small to be seen without a microscope.

Evolution predicted that animals on far islands will be related to animals on the closest mainland; that the older and more distant the island, the more distant the relationship. That's been shown numerous times to be true.

When detailed results of the Human Genome Project were announced at the 2001 meeting of the American Association for the Advancement of Science, data tied human origins to earlier forms of life, going as far back as primitive bacteria.

There's a lot more evidence showing that evolution works and creationism doesn't. The big problem seems to be no one is listening. In November 1997, a Gallup poll conducted in the United States showed that 44% of people surveyed believed that God created man less than 10,000 years ago—that evolution never happened.

## Creating the X-Men

A recent report in the *National Post* suggests that it will soon become routine for us to alter our DNA in the same "consumer fashion that people now undergo liposuction, take Prozac, and wear tinted contact lenses. In philosophical circles, this possibility is known as the perfectibility of man."[45] The report also suggests that we'll be able to buy gene products that give us perfect, soft, clear skin. In addition, we'll be able to transform ourselves from shy personalities into aggressive, dominant personalities; from wimps into risk takers; from depressed people into happy ones. For any personality factor based on genetics, researchers predict that we'll be able to alter who we are and how we react to our surroundings.[46]

[45]Everson, Brad, *National Post*, at: www.nationalpost.com/content/features/genome/0314003.html.

[46] Gresh, Lois H., *TechnoLife 2020: A Day in the World of Tomorrow*, ECW Press, 2001, pp. 165–166.

According to Dr. Henry I. Miller, a senior research fellow at Stanford University, we will think nothing of using genetic therapy to enhance our physical and mental capabilities. He points out that genetic therapy for achieving favorable physical and mental attributes is no different from going to a counselor or psychiatrist, taking psychiatric medications and anti-depressants, or getting drug treatments for baldness, obesity, and age spots.[47]

An issue of *The Futurist* predicts, "we will routinely use biotechnology to produce new strains of plants and animals by 2008."[48]

In 2001, up to thirty genetically altered babies were born and were healthy. Half of them were the result of an "experimental programme at a U.S. laboratory."[49] Genetic testing of the children established that they indeed contain genes that they didn't inherit from their parents. These children, born in 2001, are the results of germline engineering. Scientists have added genes to the children, who, in turn, will give these new genes to their offspring. Scientists have altered the fundamental genetic foundations of generations of people.

It's worth noting that researchers routinely perform germline modifications on laboratory animals, and that as early as 1990, scientists were performing somatic (non-germline, non–sex cell) genetic modifications on humans. Research articles about new experimental protocols pertaining to human germline modifications are appearing with increasing frequency.[50] The FDA is reviewing new experimental protocols about germline engineering in "increasing numbers."[51]

---

[47]Miller, Henry I., "Better Genes for Better Living," *The Wall Street Journal*. August 21, 1999, p. B1.

[48]Halal, William E., Michael D. Kull, and Ann Leffman, "Emerging Technologies: What's Ahead for 2001–2030," *The Futurist*, November 1998.

[49]Whitehouse, Dr. David, "Genetically Altered Babies Born," *BBC News Online Science Service*, news.bbc.co.uk/hi/english/sci/tech/newsid_1312000/1312708.stmI.

[50]"RAC confronts in utero gene therapy proposals," *Science*, October 1998, p. 272; see also Julie Palmer and Leroy Walters, *The Ethics of Human Gene Therapy* (New York: The Oxford University Press, 1997) pp. 85–86, 91–92.

[51]Anne M. Pilaro and Mercedes A. Serabian, "Safety Evaluation of Gene Therapies: Past, Present and Future," *American Society of Gene Therapy*, 1999, see www.fda.gov/cber/summaries/asgt060999ap.ppt.

In addition to modifying unborn children, genetic engineering has the distinct potential to create what's known as *transgenics:* the creation of embryos containing genes from other species. Specifically: "Not only can a foreign gene be put into the cells of an organism: the gene can actually be incorporated into the DNA derived from germ cells or embryonic cells of another organism. From this combination, an embryo can be produced that contains this gene that came originally from another species (called a transgene). Transgenic embryos can be put into an adult female . . . which will then give birth to [offspring] permanently carrying the transgene."[52]

The X-Men are more than possible; they're quite probable in our future.

---

[52]Rudolph, Frederick B. and Larry V. McIntire (editors), *Biotechnology: Science, Engineering, and Ethical Challenges for the 21st Century*, p. 12.

Chapter 10

# Mysteries in Space

## Science Fiction Superheroes

### Super Science without Super Heroes

**As we've seen,** the origins and powers of most major comic book superheroes were deeply rooted in science fiction. Like most good science fiction, the early superhero stories used basic scientific discoveries and advances as the building blocks for their adventures. The origins of such characters as Superman, the Flash, Spider-Man, the Hulk, and many other comic book creations were all based on science fact—even though those facts were often bent, twisted, and altered beyond recognition.

Many writers and editors in the comic book field had strong ties with science fiction, and many of them were fans of the genre long before they ever started working for comic book publishers. In the pre–World War II United States, the science fiction pulps served as the cheap paper parents of the comics.

While early comic books featured stories in a number of different fields, they became famous in America and throughout the world as the home of the superhero. Yet, that was the one area where the pulps and comics differed. Most science fiction of the period didn't feature super characters of any type. A majority of stories were about heroes, but in all except a few cases, the protagonists were fearless but otherwise ordinary humans. Even such pulp science fiction characters as

Hawk Carse and Captain Future were mere mortals gifted with extraordinary intelligence and strength. Science fiction glorified scientists and inventors as heroes, but they weren't superheroes.

Science fiction comics remained primarily superhero-oriented until the great comic book boom in the late 1940s. Looking for new areas where they could expand, comic book publishers finally tried producing comic books that featured a variety of science fiction stories that weren't about one character. The most successful of such comics were EC Comics's *Weird Science* and *Weird Fantasy* and DC Comics's *Strange Adventures* and *Mystery in Space*.

The EC science fiction comics featured stories aimed not at children but at teenagers and young adults. Most of the scripts were by writer/artist Al Feldstein and company owner Bill Gaines. They were illustrated by such talented comic book artists as Wally Wood, Roy Krenkel, Angelo Torres, and Al Williamson. The scripts were intelligently written and relied on equal parts of cynicism, sarcasm, and subtle digs at the establishment.

A number of the EC science fiction stories were somewhat revised versions of popular short SF stories from the digest and pulp magazines from a few years earlier. Ray Bradbury, reading an EC issue, spotted one of his own stories as having been rewritten for publication. He shot off a note to the publisher telling him what a great job had been done adapting the story and reminded him not to forget his royalty. An exchange of letters followed, and Bradbury, a fan of comic books for most of his life, permitted William Gaines to adapt his early horror and SF stories for EC Comics for a licensing fee of 25 cents per story.

The laws of science were rarely violated in EC comics, because they were rarely used. Super science had very little to do with most EC fiction because the spaceships and exotic planets just served as unusual locations for conventional or unconventional morality tales.

Typical was "Space-Borne," from 1952. A happy young couple take off on their honeymoon in space in a two-passenger spaceship. They go on an exploring trip and land on a distant planet. When landing, the husband, Ron, discovers his wife has a heart condition and can't go into space again. They settle on the paradiselike planet until the wife says she wants some of the comforts of civilization. Ron is forced to leave her to go back into space. His drive breaks, and it takes him six years to return. He lands and goes looking for his wife, only to be attacked by a ghastly alien monster. Ron kills it. Afterwards, he finds his wife. She's screaming in horror. When Ron left, she was pregnant. The surroundings of the strange planet changed their child into something alien. Ron had just killed his own son.[53]

EC science fiction comics were entertaining and wonderfully illustrated SF stories in the 1950s. But, there was no real science in them. EC Comics were Entertaining Comics, not Education Comics. Only a few comics tried to do both. We'll discuss them next.

## The Secrets of Other Worlds, Exposed!

Julius Schwartz, the DC comic book editor who served as one of the prime architects of the Silver Age of comics, was a longtime science fiction fan, writer, editor, and agent. In 1950, Schwartz started DC's first all–science fiction comic book, titled *Strange Adventures*. Approximately a year later, he was responsible for the launch by DC of a sister magazine for *Strange Adventures* called *Mystery in Space*. Both comics ran for years and usually featured three or four stories per issue. Some of the stories had continuing characters, but rarely did they have superheroes. The two magazines ran stories about oddball adventures involving the Space Cabbie, the Atomic Knights, Interplanetary Insurance Inc., Star Hawkins, and Knights of the Galaxy. The longest-running series in *Strange Adventures* dealt with the adventures of Captain Comet, a mental and physical marvel, who was a man born 100,000 years too early. The most popular series to run in either comic was the adventures of Adam Strange, an Earthman who teleported every month to the distant world of Rann. It was here that he and his girlfriend, Alanna, battled alien space invaders, dimen-

---

[53]Williamson, Al, "Space-Born," *Weird Science*, Nov.–Dec. 1952.

sional bandits, and numerous other interplanetary disasters that plagued the planet. Adam always managed to defeat the menaces seconds before he was mysteriously beamed back to Earth.

The stories in the two comics were aimed at pre-teens and teenagers.[54] Many of the early adventures were written by actual fiction writers, including Edmond Hamilton, H. L. Gold, and Alfred Bester. Most stories were mysteries that relied on some obscure scientific fact for the answer. Both comics often featured one-page science articles and included columns such as "Science Snaps," "Pioneers in Science," "Mysteries in Space," "Test Your Science Knowledge," and "Giants of the Telescopes."

*Strange Adventures* and *Mystery in Space* ran numerous stories about alien invaders, interstellar menaces, and heroic spacemen. Early issues of *Mystery in Space* ran a banner proclaiming "Exposing the Strange Secrets of Other Worlds!" As with many of the superhero comics of the period and to follow, often the exposé wasn't very accurate.

Take, for example, *Mystery in Space* #12, February–March 1953. The cover story advertised "The Human Magnet," a story about an alien spaceship that comes to Earth and uses a giant magnet that snatches humans into the hold of the ship. It takes a teenager to realize that the magnet isn't acting on human flesh or bone but on the metal identification bracelets the people are wearing. The major problem with the story is that while the super magnet works on the ID bracelets of scientists, it doesn't attract the metal fence circling their lab or the jeep they're riding in. Even when the young hero puts on an ID bracelet so he'll be pulled into the alien craft, the giant magnet doesn't pull the wrench he's carrying out of his pocket. Why not? That's a mystery not exposed in the issue.

Telling a complete science fiction story in six comic book pages required taking a few short cuts. Unfortunately, logic was often cut as well. That's what happened in "The Richest Man on 9 Planets," published in the same comic.

---

[54]Lois started reading these types of comics—the 1960s versions, though—when she was seven years old.

Sam Drake, a chemist working at an atomic energy research facility, feels underpaid, since all he has left of his salary after paying for his food and rent is $1. Sam, though, has invented a suspended animation drug. After depositing his $1 in the bank, Sam takes a dose of the liquid to sleep for 1,000 years. Why didn't Sam sell his formula to a gigantic pharmaceutical laboratory for millions? Obviously, that would be too simple and bring the story to a swift conclusion. Instead, Sam drinks the potion in the subbasement of the laboratory where he works. Evidently, he feels the building will remain untouched for ten centuries. Which it does. Need we even mention Sam's clothes, which are unchanged and in perfect condition after 1,000 years? Since Sam wasn't dead but sleeping, his hair should also be incredibly long. Ditto for his fingernails. But why quibble about such minor concerns?

Sam goes to the bank where he deposited his $1 and discovers it is now worth $250,000 due to the magic of compound interest, the mathematics that propel the story. No mention is made about any changes in the language over the past 1,000 years, nor does the bank manager doubt Sam's identity, since his handwriting matches the signature still on file. Worried that the value of money can change, Sam converts all his money into gold. He tells the bank executive to keep investing in gold and returns to the subbasement for several more naps. Needless to say, thousands of years later, Sam finds himself broke again when a cosmic accident makes gold the most common metal on Earth. Bad science and a bad investment strategy, Sam.

Sometimes, the errors in science were much more elementary, as was the case in "Destination—Star," published in the June–July 1953 issue of *Mystery in Space*. The opening line of the story made it clear that science wasn't going to be a major element of the story when it stated:

> Scientists have set a theoretical limit to the speed with which a spaceship can travel—186,000 miles per second . . . the speed of light. Yet, someday, if mankind is to travel to the far distant stars, it will have to travel *faster* than light.

Theoretical limit? Sorry. Even in the early 1950s, scientists knew that the speed of light was more than a theoretical limit.

The problem with too many of the stories in *Mystery in Space* and *Strange Adventures* was that they relied on humans acting stupidly or not thinking the least bit intelligently. The only method of introducing a scientific fact or curiosity was to use that information as part of the story's surprise ending—meaning that the super-scientist villains were caught by traps that wouldn't fool a twelve-year-old.

That's what happens in "The Human Fishbowl," the cover story for *Mystery in Space* #27, August-September 1955. Martians possessing super-science want Earth's mineral wealth. The Martians discover that ordinary seawater contains most of the minerals they want. The aliens trick a group of humans into thinking they are friendly, when in reality, the Martians plan to conquer the Earth. At the story's conclusion, the Martians are foiled, not by the Earthlings (who remain clueless to the invasion), but by the ocean. The invaders, we learn, land their spaceships two miles beneath the surface. When they step outside their ships, the invaders are crushed by water pressure.

In other words, the Martians, who travel in space and therefore know about atmospheric pressure in the reaches of interstellar space, don't realize that two miles of water overhead might weigh something. This implies that they really don't understand gravity, which is responsible for air and water pressure, which casts major doubts on much of their scientific expertise.

Yet, for all of the lapses in logic and scientific know-how, from time to time, the comics printed stories that weren't so easy to dismiss. Like "Doomsday on Earth," a Captain Comet adventure published in *Strange Adventures*, September 1952.

## Doomsday on Earth

The story begins with a volcano erupting in the heart of New York City, the Aurora Borealis appearing that evening in the New York suburbs, and the flow of water stopping at Niagara Falls. It continues with a gigantic whirlpool on Lake Erie and a shift in Earth's magnetic poles. Concentrating on these events, genius Captain Comet remembers reading a book that predicted similar events. It was Wilbur Walker's *The Living World*, in which the author stated that the Earth itself was a living organism and that atomic bomb tests would irritate it and cause it to react. In the climax of the story, Captain Comet, far beneath the Wyoming desert, battled a gigantic brain that was causing the disasters. Not exactly what Walker had predicted, but close.

Like most comic book science fiction of the 1950s, the story reads more like a bizarre fantasy than like anything related to science. However, the concept of the Earth being alive is no longer considered that outlandish. Proposed by the noted scientist James Lovelock, the idea is now known as the Gaia theory, the "science of the living earth."

It's doubtful that James Lovelock ever read "Doomsday on Earth." He's an eighty-two-year-old British scientist who invented the electron capture detector, a device that found the widespread residues of pesticides, leading to the publication of Rachel Carson's 1962 book, *Silent Spring*, and to the modern environmental movement. More recently, Lovelock's invention has been used to chart the global distribution of chlorofluorocarbons and nitrous oxide, two compounds that are linked with the destruction of the ozone layer in our atmosphere.

In the mid-1960s, Lovelock was asked by NASA to join a research team investigating life on Mars. He was given the job of inventing instruments to send on a Mars probe that would detect alien life. Working on the project, Lovelock decided he first needed to determine what constituted life. He reasoned that life takes in energy and matter and discards waste. Following that concept, he further decided that living organisms would use the atmosphere of their planet for this process. Humans, for example, breathe in oxygen and breathe out carbon dioxide. He concluded that life as we know it would therefore be detectable by studying the atmosphere of a planet.

Working with Dian Hitchcock, Lovelock analyzed the chemical makeup of the Martian atmosphere. The results were startling. Mars had an atmosphere that was 95% carbon dioxide, with some small amounts of oxygen and no methane. Earth's atmosphere was 77% nitrogen, 21% oxygen, and close to 2% methane. Lovelock reasoned that the atmosphere of Mars was chemically dead and that all chemical reactions occurred long ago. The Earth's atmosphere, however, was chemically active, with reactions still taking place. Considering all of this information, Lovelock theorized that the gases in Earth's atmosphere were in constant circulation, and that the mechanism driving this circulation was life itself.

Armed with this theory, Lovelock examined the history of life on Earth over the past three billion years. He used as his starting point the emergence of early bacteria and algae on Earth, living on carbon dioxide and giving off oxygen as waste. He progressed through evolution to the emergence of animals that breathed oxygen, helping maintain the atmospheric balance of the planet. Lovelock concluded that the Earth acted very much like a living organism. Helping him with his work was Lynn Margulis, an American microbiologist and expert on the role that microbes play in the Earth's soil.

Explaining his idea to the novelist William Golding, Lovelock asked for a possible name for the theory. Golding suggested Gaia, for the Greek earth goddess. In 1979, Lovelock published his book *Gaia: A New Look at Life on Earth*.

Lovelock's basic idea was that Earth was affected by the life on it: "[T]he physical and chemical condition of the Earth, of the atmosphere, and of the oceans has been and is actively made fit and comfortable by the presence of life itself."[55]

Lovelock's hypothesis was quickly embraced by some and immediately ridiculed by others. Lovelock's strongest criticism came from scientists who thought he was promoting the idea that the Earth was actually acting with a purpose. In numerous interviews for years afterward, Lovelock made it quite clear that he never once stated that the Earth had a life force that controlled the weather, the oceans, and

[55]Lovelock, James, *Gaia: A New Look at Life on Earth*, Oxford University Press, 1979, p. 15.

the lands. All Lovelock proposed was that when viewed as a whole, the Earth acted like an organism.

Still widely debated, his theory remains one of the most compelling arguments ever presented for protecting the environment. Even if the Earth is not alive, Lovelock demonstrated through model systems that major damage done to an ecosystem would be difficult, if not impossible, to repair. The Gaia theory warns that if global warming isn't addressed in the near future, then by the time life becomes unbearable, it will be too late to act. For all of our proud accomplishments, humanity is just part of a much larger environmental system, and we rely on that system to keep us alive.

Captain Comet was able to use his great mental powers to defeat the gigantic Earth brain causing worldwide disasters in one episode of his comic book adventures. If we destroy the rainforests or the life in our seas, we might discover that all the Captain Comets in the universe won't be enough to save Gaia.

## Across the Ages

Time travel—the ability to move through time in a vehicle that travels into the future as well the past—has intrigued science fiction writers for over a hundred years. The first major time travel novel, published in 1895, was H. G. Wells's *The Time Machine*. The book was less a novel about super science than a warning of class warfare. Still, Wells's novel stimulated people throughout the world. It opened the door to the future and the past, and within a few years, stories of men traveling through time were commonplace. Time travel stories were extremely popular in the pulp magazines, and some of the best appeared in the 1930s, 1940s, and 1950s. Titles included such classics as "Sidewise in Time" by Murray Leinster, "By His Bootstraps" by Robert A. Heinlein, "As Never Was" by P. Schuyler Miller, and "Seesaw" by A. E. van Vogt.

Time travel is one of the staples of science fiction comic books. In the late 1940s and early 1950s, Batman and Robin traveled in time, visiting both the past and the future in adventures. Superman was another character who was able to visit the past, most notably the planet Krypton before it exploded. The Fantastic Four visited ancient Egypt, and the Avengers, another superhero group, visited the far distant future. These stories were extremely popular, as were the time travel stories that appeared in the science fiction comics of the 1950s like *Strange Adventures*, *Weird Science*, and *Mystery in Space*, mostly because they used unusual plots not possible in any other type of story.

For example, consider "Across the Ages!," the cover story for the September 1955 issue of *Strange Adventures*. In the future, the National Historical Society features lectures by great figures from history like Julius Caesar. Time travel agents bring the notable into the future from the past to speak to the society. When the lectures are finished, the agent returns the speaker to approximately the same moment he or she disappeared. Before releasing the famous person, the agent wipes clear any memory of the future, ensuring that no paradoxes arise from the time trip.

However, sometimes the best-laid plans of time travel agents go wrong. On his latest mission, the nameless Historical Society agent has to bring back Cleopatra, Christopher Columbus, and Napoleon. All three have tremendous egos and argue who is the most famous.

Something unexpected happens to the motor in the time machine, and the device comes to a halt in the twentieth century in a museum. Cleopatra finds a statue of herself but is annoyed by its inaccuracies. Napoleon and Columbus go into the city to see who is more famous. Columbus stumbles on a Columbus Day parade, while Napoleon finds a whole section of the local library devoted to works about him. In the end, all three time travelers return to the repaired time machine and take off for the future.

It's a short but pleasant story. We can't help but notice the similarity of "Across the Ages" with the movie *Bill and Ted's Excellent Adventure*, where two high-school students go into the past to locate famous historical figures to help them with their term paper.

In "Assignment in Eternity," published in *Strange Adventures* for August 1957, a nameless college professor is startled one day when Abraham Lincoln and George Washington march into his classroom. The two turn out to be drama students in makeup for a play. Still, their appearance triggers some odd thoughts in the teacher's mind. Working alone at home, he scribbles a formula proving that time travel actually works. The professor soon realizes that he's a time traveler from the future sent back to Earth to save his time from a dreadful catastrophe. The story concludes with the time traveler remembering his mission and saving the future.

Again, it's a low-key story, without any violence and with very little action. It's a puzzle story in which the nameless hero needs to sort through data logically to discover the answer to a problem. Both *Strange Adventure* stories start with a basic assumption that time travel is possible, an idea considered totally impossible during the 1950s. Now, studies based on quantum mechanics and Einstein's general theory of relativity have made it clear that time travel might not be as impossible as we once thought.

In the 1930s, Albert Einstein, working with Nathan Rosen, concluded that the general theory of relativity implied that black holes served as a gateway between two regions of space. This gateway was called the "Einstein-Rosen bridge." Most people felt that these gateways were imaginary exercises dealing with impossible objects (black holes).

In the 1960s, the mathematician Roy Kerr did a mathematical analysis of black holes as predicted by the general theory of relativity. At the time, black holes (as discussed in our chapter about the Green Lantern) were considered possible, but no one actually believed they existed. Kerr discovered that if a black hole was rotating, the singularity formed would be in the shape of a donut instead of a mere point. According to Kerr's calculations, it would actually be possible for a spaceship to pass beyond the event horizon of this rotating black hole and travel through the space inside the singularity ring. This Einstein-Rosen bridge would then transport the spaceship through hyperspace to another location. Kerr also noted that the gateway would work through space and time, since Einstein's theory dealt with both.

Still, Kerr's theory dealt with rotating black holes, and, at the time, everyone felt certain that real black holes didn't exist. Few mathematicians or astronomers paid much attention to Kerr's work until the 1970s, when astronomers began discovering what seemed to be real black holes.

With black holes no longer imaginary, Einstein-Rosen bridges and Roy Kerr's work took on new meaning. Scientists began studying these black hole gateways connected to other black hole gateways, a concept that became known as wormholes.

However, it wasn't until the astronomer and writer Carl Sagan signed a contract to write a novel about man's first encounter with extraterrestrial life that time travel theory really exploded. Sagan, who planned to use wormholes as a method of interstellar travel in his book, *Contact*, wanted his story to be as accurate as possible. So he asked Kip Thorne, a professor at CalTech, for realistic details. Thorne, an expert on general relativity, was intrigued enough by Sagan's concepts that he and several graduate students began studying black holes and travel through space and time.

Much to Thorne's own surprise, he and his assistants discovered that nothing in the general theory of relativity prevented time travel. Traveling through a wormhole, a spaceship could travel backwards or forwards in time. There was only one minor problem. Anyone wanting to construct a time machine needed scientific knowledge about working with black holes: not impossible, but not foreseeable anytime in the near future.

During the past fifteen years, time travel has been examined by scientists and mathematicians all over the world. It's been declared impossible, but no one's found any credible evidence that it can't take place. As far as we can tell in 2002, time travel may be possible. There's only one challenge that makes any sense. It's a problem less of mathematics than of logic. It has no formal name, but it's often called the "grandfather paradox."

## The Grandfather Paradox

In "Even the Heir," published in *Weird Science* #16, November–December 1952, the main character, Seymour, mixes time travel with murder in a bizarre scheme to get rich quick. This being an EC story, things don't turn out the way Seymour planned, and in the end he's left literally with nothing. Still, the concept of the story gives us plenty to discuss.

The grandfather paradox is one of the most popular themes in time travel science fiction. It's also the one challenge to the possibility of real time travel. And, strangely enough, the one answer that seems to answer the dilemma of the grandfather paradox is right out of science fiction.

Simply put, the grandfather paradox asks: What would happen if a man built a time machine, traveled back in time, and killed his grandfather before he got married?

The simplest answer, the one used in "Even the Heir," is that the person would cease to exist. If he went back in time and killed his grandfather before he married, then his father (or mother) wouldn't have been born. Needless to say, without his father being born, he'd never be born. By killing his grandfather, the murderer would have been committing suicide—or would he?

If our murderer killed his grandfather before he got married, then the killer would never exist. If he never existed, then how can he do anything? Not existing means he never did anything, so his grandfather was never killed. But if his grandfather never was killed, then he did get married, our killer was born, and his grandfather was killed.

In short, building a time machine and traveling back in time creates a paradox. Time paradoxes are quite annoying because they don't have simple solutions. In the hands of an expert writer and logician, time paradoxes are maddening. What is perhaps the greatest time travel story ever written, "All You Zombies" by Robert A. Heinlein, stars a character, Jane, who, through misadventures in time, ends up as *both* her own father and mother, as well as all the rest of the characters in the story.

Many scientists feel that the grandfather paradox and similar time travel paradoxes clearly demonstrate that time travel is impossible. However, there's at least one fairly controversial theory that says otherwise. Proposed in 1957 by Hugh Everett III and more recently championed by David Deutsch, this theory offers a simple alternative to time travel paradoxes. It's known as the Many-Worlds Interpretation of Quantum Mechanics.

According to Everett's theory, whenever multiple possibilities exist in quantum events, the world (by "world" we mean the entire universe) splits into *many* worlds, one for each possibility. Each of these worlds or realities is as real as the original and exists simultaneously with the first, while remaining unobservable by any of the others.

In simple terms, the Many-Worlds theory argues that for every possible outcome of every decision ever made, an alternate reality exists. These alternate realities form what Everett called the multiverse. For example, while reading this chapter you might decide to have a snack. You look in a cabinet and find popcorn and potato chips. You take the popcorn. That's one reality. At the same instant, another reality comes into existence in which you took the potato chips instead. A third reality also is created in which you take both popcorn and potato chips. And, needless to say, a fourth reality emerges in which you change your mind and decide not to eat a snack.

On a larger scale, a reality exists in which Germany won the Second World War. Another reality exists in which Hitler was never born. A reality exists where the United States never entered the war. Realities for every possible branching point in history, no matter how minor the decision or how major the decision, exist. With each reality branching just like the first, every decision made in the alternate world leads to more alternate worlds. Everett's multiverse spreads like the branches of an enormous tree. It's a cosmology in which everything that can happen happens somewhere.

How does Everett's multiverse tie in with time travel paradoxes? Simple. The Many-Worlds theory postulates that whenever an event is changed, a new reality is created. If a past event is changed by a time traveler, no paradox arises, because the time traveler finds himself in a new reality that his actions just created. He's left his original reality and entered a new one.

For example, in the grandfather paradox, a man goes back in time and kills his grandfather. In a one-world reality, this murder creates a paradox. In a Many-Worlds reality, a new reality is created by the murder. It becomes a reality in which the killer was never born. Our killer has passed from one reality to another, a reality that he created. If he travels to the year he came from, he'll find no trace of his existence. But it doesn't matter; because he didn't actually come from this universe, but from an alternate one. He's caused his reality to branch, and thus there's no paradox possible.

The Many-Worlds theory is controversial and isn't accepted by many physicists. But none of it contradicts any of the basic laws of the universe. If anything, the Many-Worlds theory explains some of the more troublesome problems still existing with quantum theory. It's a fascinating idea, especially since it's been a staple of science fiction since the first parallel universe story was written in 1919. It's also the stuff of science fiction comics. Perhaps the very best alternate world comic book story was published in *Weird Fantasy* #20, July–August 1953, titled "—For Us the Living." It told the story of an innocent tourist from the future who wanted to see the greatest statesman in history, but accidentally sets off the string of events leading to the murder of Abraham Lincoln. It's a perfect demonstration of how even the best intentions can cause terrible results when combined with time travel and alternate realities.

Most of all, the story serves as a perfect example of how science and science fiction, characters and plot, joined with great artwork and a sense of style can tell a great story in an entertaining manner, which is what comic books are all about.

Chapter 11

# The Right Stuff

## Donald Duck

### The Real Deal

**Superhero comics inevitably** involve some sort of science or super science. But comic book writers walk a fine line between what's possible and what's not. Often that line fades or disappears in light of what needs to happen in the story, and scientific accuracy is sacrificed to keep the plot moving. The excuse is always: "It's only a comic book."

Comic book fans have always complained about the genre's lack of respect. They feel comics deserve more attention than they receive in the media and that the general public should realize that comics aren't just for kids. The success of a few graphic novels in the mainstream has spurred hopes that someday comics will achieve some measure of respectability. Maybe. Someday. When comic book writers give the same care and attention to all story elements—including the science—that writers of other forms of literature apply to their work.

It is possible. It has been done before.

One comic book writer never cheated his audience. The author of over five hundred comic book stories, he used real science and real technology in his stories. When characters visited a foreign country, this writer made sure his backgrounds were authentic. Best of all, although he aimed his comics toward twelve- and thirteen-year-olds, he still made sure there was enough rattle and clack on the train to convince the most gung-ho general that a war was going on outside his apartment. He was a modest soul whose genius won him millions of fans worldwide.

His name was Carl Barks. And the names of his superheroes were Donald Duck and Uncle Scrooge.

## The Duck Man

For approximately twenty-five years, from 1942 to 1966, one man wrote and drew nearly all of the Donald Duck comic book adventures published in Disney comics. The same man invented Donald's Uncle Scrooge and a host of other supporting characters, including Gladstone Gander, the Beagle Boys, and the Junior Woodchucks. All of this work, over five hundred comic book stories, appeared anonymously, but readers all over the world knew it was the work of the same artist. It was obvious from his art, from his characters, and from his incredible storytelling skills. When another artist did a fill-in story for *Walt Disney's Comics and Stories*, letters would immediately pour in, demanding the return of the "Good" Duck Man. That writer and artist was Carl Barks.

Carl Barks was born in 1901 in Canada. His family moved from place to place as his father looked for work, and Barks didn't finish grade school until he was fifteen. He never attended high school. After years of hard physical labor on a farm, riveting on a railroad, logging, and worse, Barks became a cartoonist in the early 1930s. He started working for the Walt Disney Studios in 1935, and he began writing *Donald Duck* comic books in 1942. Uncle Scrooge was created as a supporting character for Donald Duck in 1947, but by the early 1950s he had become so popular that he merited his own comic. By the mid-1950s, *Uncle Scrooge* comics were selling three million copies per issue. Barks retired from the comic book field in 1967. In 1984, U.C.L.A. Professor Geoffrey Blum said of Barks that he "may well be the most popular writer in America today—certainly he is one of a handful."[56] Carl Barks died on August 25, 2000, at the age of 99.

Barks was known not only for the great humor in his work but also for his great storytelling ability. In a tribute to Barks, George Lucas wrote, "I think the reason Carl Barks's stories have endured and have had such international appeal is primarily their strength as *good stories*."[57] The noted science fiction and fantasy author Alan Dean Foster took this idea one step further in an essay about Barks, in which he wrote, "In the 1940s and 50s . . . the most fortunate children had private tutors. Those who were even luckier had as their tutor Carl Barks, who provided an introduction to many advanced subjects in the guise of illustrated entertainment."[58]

Carl Barks never underestimated the intelligence of his audience. In an interview with Barks published in 1985, he stated, "I didn't go along with the editors' idea that children who bought . . . comics were ignorant, blubbering infants. I assumed that my average reader was around twelve years old, semi-worldly, and already quite knowledgeable about mechanics, history, science, nature, travel, and so on."[59] Barks also said, "I felt that readers of any age would be more pleasantly entertained by stories that were plausibly written and believably drawn."[60]

Examples of Barks's casual use of science and technology are easy to find. In "Under the Polar Ice," *Walt Disney's Comics & Stories* #232, January 1960, Donald and his three nephews travel by submarine under the arctic ice cap. The submarine drawn by Barks, as well as the interior of the ship, is based on actual photographs of the atomic

[56]Blum, Geoffrey, "Introduction" *The Carl Barks Library of Walt Disney's Uncle Scrooge 1–20*, Another Rainbow Publishing, Scottsdale, AZ, 1984.

[57]Lucas, George, "An Appreciation," *Uncle Scrooge McDuck, His Life & Times*, Celestial Arts, Millrae, CA, 1981.

[58]Foster, Alan Dean, "The University of Barks," *The Carl Barks Library of Walt Disney's Uncle Scrooge 1958–1963*, Scottsville, AZ, 1985, p. 221.

[59]Ibid.

[60]Ibid.

submarine USS *Nautilus*, which went on the same trip a little more than a year before. In *Uncle Scrooge* #46, December 1963, Uncle Scrooge and Donald descend six miles beneath the sea in a bathyscaph based on the design of the *Trieste*, an actual bathyscaph that had gone nearly seven miles into the Pacific Ocean three years earlier. In the same story, a tank of gasoline is kept on top of the bathyscaph to provide the necessary buoyancy to take the ship up (as gasoline is lighter than water). Sonar is used underwater to keep track of where they are going, and when air runs short, Uncle Scrooge talks about fantasies brought about by carbon dioxide poisoning.

In "Vacation Time," published in *Vacation Parade* #1, July 1950, Donald Duck and his three nephews go on a camping trip to the mountains. Donald and the boys are good campers and make a safe fire inside a ring of rocks. Not so careful is a tough, inconsiderate camper who smokes cigarettes and doesn't know how to make a safe fire. Needless to say, the tough guy's fire starts a forest fire. Donald and his three nephews are trapped in the wilds by the huge blaze. One of the nephews suggests jumping in the river but is told by the others that he would be boiled alive. Donald instead has the boys take their canteens and shovels, and he searches for a clearing.

In the clearing, Donald has the boys dig trenches deep enough to lie in. The ducks wet their coats with canteen water then lie down in the trenches with the wet jackets over their heads and their shovel blades keeping their wet jackets up to give them air to breathe. Then Donald covers his nephews with dirt, and goes through the same procedure for himself. The forest fire roars over the ducks but they remain safe. In the end, the tough guy is caught by the forest rangers for starting the fire and Donald and his nephews enjoy their vacation, having been saved by good science.

One of the most famous Uncle Scrooge adventures, "The Flying Dutchman," *Uncle Scrooge Comics* #24, March–May 1959, is also one of the most interesting. Uncle Scrooge, with Donald and his nephews in tow, goes looking for a long-lost missing merchant ship that turns out to be the famous ghost vessel the *Fliegende Hollander*, the Flying

Dutchman. According to legends, the boat sails high in the sky on stormy nights with red sails, and always travels against the wind.

Barks carefully gives his readers a full account of the legend when the ship first appears, giving it a supernatural spin. He then uses science later in the story to explain away all the supernatural details. The haunted ship turns out to be a merchant ship deserted by its crew because of plague and then frozen in the Antarctic ice. "Pressure ridges from pack ice," declares one of the nephews. "The aurora and the phosphorus bounced its image between the clouds and the sea for hundreds of miles," explains another nephew. The final mystery of the ship is revealed at story's end, when Scrooge and company see the iceberg swing about to point into the wind, thus giving the illusion that the boat is sailing against the wind. Could science be any more entertaining?

Perhaps the most amazing science feat of all of Carl Barks's duck stories started with the lead story for *Walt Disney's Comics and Stories* #104, May 1949. In the ten-page adventure, Uncle Scrooge and Donald get into an argument about raising a sunken ship from the bottom of the sea. Working on a tight budget, Donald and his nephews fill the inside of the boat with Ping-Pong balls. The Ping-Pong balls fill the hold, displacing the water in the ship and giving it enough buoyancy to float to the surface. At the time, the story drew little notice.

Things changed in 1964 when a Danish scientist, Karl Kroeyer, raised a sunken freighter near the Kuwait harbor in the Persian Gulf. Kroeyer raised the ship by pumping it full of expandable polystyrene foam bubbles, which worked in the same manner as the Ping-Pong balls in the Donald Duck adventure. Where did Kroeyer get his idea? *From reading the comic book story as a young boy in Copenhagen!*

*Later, when Kroeyer tried to patent his process, he was denied his request because the method had been published in the comic book fifteen years earlier!*

Comic books and science do mix. It just takes a genius like Carl Barks to make them mix well.

## Appendix A

# Who Missed the Cut?

**Reading this book,** someone who doesn't know much about the comic book industry might conclude that there are only two comic book publishers, DC and Marvel. Nothing could be further from the truth. Every month in the United States, approximately 300 comic books are published. About one-third of these are published by the two giants of the industry, DC and Marvel (known as the Big Two), but the other 200 or so are produced by dozens of smaller companies located throughout the United States and Canada. These other comics are read and collected by a thriving audience that ranges in age from eight to eighty. Some of the more successful smaller publishers in the comic book field include Image Comics, Top Cow, Crossgen, Chaos Comics, Black Bull, Humanoid Publishers, and Oni Press. There are many others.

In the early 1940s, when Superman and Batman were the big kids on the block for DC, there was plenty of competition from publishers such as Fawcett, Fiction House, Quality, and Timely comics. During the more than half century since the first appearance of Superman, there have probably been well over 1,000 different comic book titles published, with half of them being superhero comics. That's a lot of superheroes. They're obviously not all covered in this book. Why not?

First and foremost, there's not enough room in this book to talk about all the superheroes. We included as many different superheroes as we could without cutting down the space necessary to discuss each character and his powers in the detail we felt he deserved. In this, we followed Gauss's motto, "Pauca sed matura."[61]

We didn't have enough room to cover every important superhero who appeared in Marvel and DC comics during the past few decades. Thus, we weren't able to discuss Captain America and genetic engineering, or the Vision and androids, or Wonder Woman and her invisible airplane. Nor did we have the space to examine the many wonderful supervillains fought by our superheroes. There's no Lex Luthor, Mr. Freeze, the Mole Man, or Dr. Octopus in this book, although we love them and the manner in which they twist science for their own despicable goals. For them, hopefully, there will be another time.[62]

A number of popular characters from the Big Two publishers weren't included in this book because of their supernatural origin or powers. Thus, Marvel Comics's *The Mighty Thor* didn't make the final cut; nor did DC Comics's *The Spectre*. Science has little to say about the supernatural, other than that it's impossible.

We originally planned to devote a chapter or more to the superhero women of comics. But reviewing the history of women in comics, we came to the same conclusion reached by most historians of the field—that women never really received the attention they deserve in comic books. As of 2002, they're still waiting.

Comic books, like most pop culture creations printed before the 1960s, were incredibly sexist in their portrayal of women. The only truly powerful and intelligent women of the period were the femmes fatales who regularly tried without success to seduce our heroes away from the straight and narrow. Leading that small but select group was Selina Kyle, aka Catwoman, who battled Batman in several memorable encounters.

As society's views of women changed, so did women's place in comics. In the past thirty years, women became more visible in the world of superhero comics. But unfortunately, that visibility wasn't

[61]"Few but ripe."

[62]Send a postcard to our publisher demanding a followup book!

only as characters but also in regard to their costumes. Empowered women seemed too often to mean unclad women. The new breed of women superheroes and supervillains wore painted-on, skintight bodysuits or outfits straight out of swimsuit catalogs—or less.

There are more women superheroes now than at any other period in comic book history. There are female assassins, ninjas, and detectives. There are women with magical powers. But there still aren't many females with scientific backgrounds or super powers who aren't grounded in fantasy and magic. Too many female lead characters are still just beautiful bimbos in thongs. In the minds of comic book writers and artists, women and science don't seem to match. Hopefully, in years to come, that will change. We can only hope.

What about the non–DC or Marvel comic superheroes? In the 1940s, too many of them were played for laughs and had no ties to logic, much less science (Plastic Man, for example). Others from the period had supernatural origins (Captain Marvel), or they had no super powers at all, but just seemed to like running around with a cloak and hat, fighting crooks (the Green Hornet).

In the 1950s and 1960s, DC and later DC and Marvel had a near-monopoly on superheroes. The only other notable comics being published were from Disney and Archie. For a short period, *Archie* comics published a line of superhero comics. They're better left forgotten.

In the 1970s, the competition heated up as a number of independent publishers once again realized that comics could be profitable. The market kept on expanding until the 1990s, when a glut of titles and the collapse of the collectors' market saw a sudden crash in the comic book field. The second half of the nineties was a period of retrenchment and regrouping.

Now, in the first few years of the new century, comic book sales have started to increase, and more titles are being published every month. Several well-financed small publishers have entered the market with aggressive plans for promoting and selling new titles with all new characters. Comics seem to be making a comeback, as a new wave of expansion ripples through the industry. Unfortunately, the level of science and technology in these books hasn't risen with the same tide.

In preparing this appendix, we surveyed the fifty most popular titles not published by the Big Two over the past ten years. A number of the comics, such as *Spawn*, *Witchblade*, *Sandman*, and *Vampirella*, featured characters involved with the supernatural, not science. The rest, though supposedly based on advanced science and/or technology, had no relationship or connection with either. Heroes were the result of genetic manipulation or atomic research, or were mutants. The explanations and origins of these characters were minimal, and their powers remained unexplained, illogical, and, in most cases, totally impossible. Action was clearly the winner over reason, style over substance. Again, we hope this will change. Only time will tell.

## Appendix B

# The Professionals Speak

**It seems only** proper to conclude this book with some words from the people who made this book possible—the comic book creators. A group of questions dealing with science in comics was sent to a number of top professionals working today in the comic book field. What follows is something about each creator, and then their answers to our questions.

**LEN WEIN** is one of the most creative people ever to work in the comic book field. He's worked as a writer, an editor, a cover editor, a senior editor (DC Comics), an editor-in-chief (Marvel Comics and Disney Comics), a colorist, and a penciller. He was the writer/co-creator of *Swamp Thing*, *The Human Target*, *Wolverine*, the *New X-Men*, *Gunfire*, and many other characters. He's written *Superman*, *Batman*, *Wonder Woman*, *Green Lantern*, *Justice League of America*, *Phantom Stranger*, *Blue Beetle*, *The Amazing Spider-Man*, *The Incredible Hulk*, *The Mighty Thor*, *The Fantastic Four*, *The Defenders*, *Star Trek*, *Star Wars*, *The Victorian*, and countless others.

Len's won the Shazam Award for Best Writer, Best Story, and Best Continuing Series; the CBG Award for Best Editor; the GEM Award for Best New Company; and many others.

Len's current comic projects include *The Victorian* (an ongoing monthly series from Penny-Farthing Press), and *Batman: Nevermore* (five-issue Elseworlds mini-series).

**MARK WHEATLEY** has done it all. As Mark puts it, "I started working in a newsstand, went on to printing, and then to art directing. I've created, drawn, written, lettered, painted, colored, edited—you name it. Currently I am the publisher of Insight Studios Group. I work in creative credits whenever the opportunity arises. But, time is tight, since the publishing is booming for us these days."

Mark is known internationally as an accomplished illustrator, writer, editor, and publisher. He's won the Inkpot, Speakeasy, Gem, and Mucker Awards and has been nominated for the Harvey and Ignatz Awards for his comic book and pulp creations, which include *Breather, Radical Dreamer, Mars, Black Hood, Tarzan, The Flash, Prince Nightmare, Blood of the Innocent,* and *Titanic Tales.* His illustration work, chosen for inclusion in the annual *Spectrum* selection of the best in fantasy and science fiction art, has also appeared in magazines, books, comic books, and games. He has written books, comic books, and television shows. In the last few years he has established himself as a premier designer and editor of high-quality art books, which include the record-breaking *Frank Cho Illustrator, IS Art: The Art of Insight Studios,* and *Gray Morrow Visionary.*

Currently Mark is writing the popular *Hammer of the Gods* comic book, which was created and illustrated by Michael Avon Oeming. A new series of *Hammer of the Gods* stories will be appearing in the summer of 2002. Mark is also editing a collection of Al Williamson art and stories for summer 2002 with the title of *Al Williamson Adventures.* A second volume of the pulp revival *Titanic Tales,* also in production and scheduled for 2002 release, will feature Mark's whimsical horror creation, the Frankenstein Mobster.

**BRETT BOOTH** is a superbly talented comic book penciller (artist) and writer. Brett's many credits include work done on *The Kindred 1* and 2 mini-series, *Wildcore* 1–6, *Backlash, The Fantastic Four Heroes Reborn* 6–9 and 11–12, *X-Men Unlimited* 25–29, *Backlash/Spider-Man,* and *Extinction Event.* Upcoming work by Brett includes *Kindred 2* and more issues of *Extinction Event.*

**BUDDY SCALERA** is a writer, filmmaker, journalist, editor, and web designer. As a comic writer, he's best known for his work on *Dead-*

*pool*, *X-Men Unlimited*, *Necrotic*, *Decoy*, *Elvira*, and *Desperate Times*. As a journalist, he's written for *Wizard*, *Comics Value Monthly*, and the *Comic Buyers Guide*.

Buddy was Wizard Entertainment's first on-line editor, developing wizardworld.com. He also helped design wizardschool.com, toyfare.com, inquestmag.com, and a number of other web sites. At present, he's developing a new web site called comicbook school.com, an online school for aspiring comic professionals. He's also working on *Weapon X: The Draft*, *Necrotic: Book 2*, *Decoy*, and several other comic book projects.

**MORT CASTLE** calls himself a freelance comic booker: "I've written, edited, packaged, created, marketed, hyped . . . One of these days, I'll find some area in which I am successful."

Mort created *Horror: The Illustrated Book of Fears*, *Dream Angel*, and others; produced and packaged with/for *Innovation* J. N. Williamson's graphic novel series *Masques*, with such writers as Stephen King, Robert R. McCammon, and Bob Weinberg, and artists Olivia, Boris Vallejo, and Mark Evans; wrote the four-issue adaptation/spinoff of *Leatherface—Texas Chainsaw Massacre III*; wrote *Monolith*, for Comico, The Comics Company ("as intelligent a superhero series as I could make it"); created and wrote the one-issue series (?!?) *Night City*, with Mark Nelson and Don Kramer as artists. It was the 1998 International Horror Guild award nominee for best graphic story/collection and it "gave a start to many people who have since done considerably better in this biz than I."

At the moment, Mort's working outside the comic book field but writes, "I imagine I'll soon be hearing from Eclipse, Kitchen Sink, or First Comics."

**MAX ALLAN COLLINS** is one of the most popular mystery writers in America today. A longtime fan of comic books and comic strips, he has written both.

As a mystery writer, he's a two-time winner of the Shamus Award for the best novel for the Nathan Heller historical mysteries. His Ms. Tree work (co-created with artist Terry Beatty), the longest-running private eye comic book, has been nominated for the Eisner Award. He's also known for his innovative early use (1981) of the tough female P.I., predating series by Sara Paretsky and Sue Grafton.

Some of Max's many other credits include successor to Chester Gould as writer of the Dick Tracy comic strip 1977–1993, writer of several Batman projects—a year run of the comic book, a graphic novel, the first story, and writer and cocreator of Wild Dog and Mike Danger.

Max created the graphic novel *Road to Perdition* (1998; with artist Richard Piers Rayner), which was adapted into a DreamWorks film starring Tom Hanks and Paul Newman, directed by Sam Mendes.

The writer/director of three independent films, most recently Max is discussing a new Batman project with Andy Helfer at DC. *The Road to Perdition* is being reprinted in conjunction with the movie. Max isn't working on any other comic projects at the moment, as he's concentrating on novels and screenwriting (and directing indie projects).

**Question 1: Do you think real science has a place in today's comics?**

LEN: Yes, but only when it doesn't get in the way of telling our stories the way we want.

MARK: Science is the revealing of the underpinnings of reality. An understanding of science is a must for creating convincing science fiction. It makes no difference if the story is told in text or in comics—the science should be there. Otherwise it just becomes fantasy. I've written plenty of fantasy. But when it comes to SF—I've always done my research and carried my speculation from that grounding. *Mars* was my first science fiction comic book series and was solidly based on research into the nature of the brain/mind connection—as well as the biomechanical potentials of reestablishing the severed links between spinal nerves in paraplegics. We also were in touch with NASA for the latest information on the Mars landings. On the other hand, not much of that research made it onto the printed page in the comic books. The stories had to be about people and their reactions to their surroundings and each other. The science was there, but hidden under the surface. Since the first publication of *Mars* in the mid-1980s, walking paraplegics have become a reality—and the systems used greatly resemble what we predicted.

BRETT: Yes, there will always be an element of disbelief, but if proper science can be used to give the book a more realistic feel, then go for it. Getting the science right in the first place will prevent you from having to come up with some highly implausible science at a later date.

BUDDY: Yes, absolutely. I believe as science goes pop culture through TV channels like Discovery, the reading public becomes more savvy about details and facts. They are willing to temporarily suspend disbelief on certain aspects of a story, but I do believe that they truly yearn for something concrete and tangible with their sciences. If they have been properly educated about the facts, it allows them to relate to the story on a deeper level.

MORT: Yes, for those few writers who dig science. Not so long back, there was an entire comic book universe based on principles of "quantum mechanics"; most of the writers' knowledge of the subject was limited to how to spell "quantum."

MAX: Science always has a place in serious science fiction; but fantasy, even mixed with SF, can, and even should, play by rules of the writer's invention.

**Question 2: Do you think there is more or less science in comics than there was in the 1960s (or any other period of comic book history)?**

LEN: Less.

MARK: Science was the core of many of the stories concocted for the DC comics titles of the 1960s. *The Atom*, *Green Lantern*, *The Flash*, and *The Justice League of America* were all written by SF authors and their background showed. Stories were centered on a bit of scientific trivia. They were often clever, structured with the same kind of payoff you might earn from a riddle. But they were usually flat when it came to emotional hooks or conflict. I enjoyed them when I was young, but by my teen years I was much more interested in the hotter emotional content of the Marvel titles. And the Marvel titles had very little accurate science. That didn't stop me from capitalizing on my comic book science knowledge. In the seventh grade I remember impressing my science teacher by having the correct answer to his question about alternate matter. "Is there anything other than matter in the universe?" "Anti-matter!" I told him. He launched on a lecture about how it was so important that we do outside reading using me as a good example until he asked me where I'd learned about anti-matter. The laughs of the students about my reading about anti-matter in an issue of the *Justice League* could have pulled the rug out from under the teacher—but he was smart enough to take it in stride. He recommended that we all start reading comic books if they were full of useful information.

BRETT: Less—most of the comic "science" was established in the '60s. What writers do now is just build on to what was originally done.

BUDDY: I think that these days there is more concrete science than ever. Yet, I think the science tends to be more below the surface. Rather than being the theme, it is part of the undercurrent. I think the 1960s had comic books that were fueled by a basic misunderstanding of real science. For example, take this line from *The Fantastic Four* #1:

> SPX: Rak Tac Tac Tac Tac
>
> BEN: Hear that?? It's the cosmic RAYS!! I—I warned you about 'em!!

It's reminiscent of duck-and-cover drills of the 1950s. Now that we know about radiation, we know that wouldn't have saved anyone, unless their moms dressed them in lead underwear that morning.

MORT: There "might" be more pseudo-science—and it is faked better now that every kid has TLC and the Discovery Channel. But if we want sciencey-science, we don't go to any form of popular entertainment. *Jurassic Park*? It wasn't about cloning. It was about dinosaurs stomping ass.

MAX: The SF craze of the '50s probably led to more invocation of science than in any other era—though the accuracy was spotty, to say the least.

**Question 3: Do you do any research on science when writing superhero comics that are heavily involved in science or technology? Do you feel that the needs of the story outweigh the reality of the science?**

LEN: The story always outweighs the science.

MARK: I'll send you back to my answer for #1.

BRETT: Yes, I don't like to write about something if I have no idea how it works. Sometimes the story does outweigh the science; the trick is getting the science in regardless.

BUDDY: The needs of the story are always going to outweigh the scientific evidence. But I research history when I am going to do a historically based story, and science research when I am writing a futuristic tale. Yet, I am generally looking for something to support my existing ideas, more than trying to discover new scientific or historic concepts. So the idea usually comes first, then the science that may support it. If it does not support it, well, hey . . . "must be alien technology."

MORT: There was a lot of solid scientific knowledge that went into my work on Leatherface—most of it dealing with mortuary and forensic sciences. You do the research you have to do to create an enveloping and convincing world and that is that.

MAX: When I'm writing a science fiction project—Mike Danger, for example—I try to keep it as real as I can. I'm a research fiend, but science is my weakest area . . . history is my strength. Fortunately, my son Nate (a computer science major going to college) is everything I'm not in that area—and he's my in-family advisor.

The science should be fundamentally solid—people with weak science background like me don't have a story spoiled by inaccuracies, but those who have solid science knowledge will be pulled out of the fiction's reality. That's poor storytelling. But questioning the science in a nitpicky way—that's too much. The tail should not wag the dog. In my historical fiction, for example, I try very hard to maintain accuracy; but if the story needs for me to compress the action that took place in a period of one month into a week or even a single day—in order to make the story more effective—I will do it without shame.

**Question 4: Sometimes superhero comics use ideas that are improbable but not answered (like life on other planets). Other times, comics use ideas that have been proven impossible by science (like moving faster than the speed of light). Do you think most readers know the difference between the two? How about most writers?**

LEN: Most readers, I doubt it. Most writers, I'd like to think so.

MARK: I've noticed that the readers I meet at conventions, fans of my SF stories for *Mars* or *Radical Dreamer* are usually very intelligent! They often know more about the science involved than I do.

BRETT: Hopefully! But not everyone does. In fact I just saw something that showed some type of wave (micro?) moving faster than light, so it is possible. I don't think comic writers are always up on all the current science, so a little research should always be done.

BUDDY: No and maybe. I believe that as students of science, our audience tends to be on the youngish/male side. That demographic does not gravitate toward science reading as much as it just wants really cool stories. They want ray guns and blasters, even if they never really stop to think if they could work or not. As long as it looks cool, it works for them. But you give them a piece of science that they know cannot possibly work, you lose them. (Exception: Unless you contrive a really cool explanation.)

And comic writers run the gamut, so I would have to say maybe. I think that they know the difference, but they also know the rules. The rules allow you to utilize anything that has been done before. (Unless continuity has been revised.) With Cable, for example, you could blink the character through time. This isn't something that's been scientifically proven, but you could do it because someone else has done it before. Now, we know that Mr. Fantastic can stretch. But if Cable were to begin stretching, fans would cry foul. Unless, of course, you had a really good explanation.

MORT: No, most readers don't know the difference. Most writers don't know the difference. Most of the writers who present martial arts scenes in their comics know that after their extensive research into the subject, there probably is a "horse stance" in tae kwan do, or gung fu, maybe, or chow mein . . . and most readers know that Bruce Lee was better looking than Jackie Chan, but Chan, being alive, is probably the superior martial artist.

MAX: Enough do to be careful about it. But superhero comics tend to be more fantasy than science fiction—even though Superman has certain SF trappings, the concept has much more to do with childhood wish fulfillment than science.

That's why I prefer—when I'm given the chance to do mainstream comics—to write Batman. Wild as it is, the Batman concept is a fantasy whose trappings are real-world . . . noir, crime novel stuff. The problem with superheroes is that the parameters keep shifting—Superman has such-and-such a power because he needs it at the moment. The worst sin Batman commits is conveniently packing in his utility belt the gizmo he'll happen to need on a given crime-fighting night.

**Question 5: As our world becomes more technology-oriented, do you see science becoming more or less important in comics?**

LEN: Less.

MARK: I think science has a place in the stories that entertain our culture. We tell ourselves stories in an effort to understand our world. Research into science has much the same motivations—the need to understand. The more we need to understand how science affects our daily lives, then the more we will see stories in books, comics, movies, and television that will deal with the subject. Hopefully, as creators, we will get the science right!

BRETT: Hmmm, I'd like to say less. We need more fantasy-oriented comics. But with new science come new ideas.

BUDDY: More important. Look how many superheroes check their e-mail. And in *Deadpool*, we would purposely write in little gadgets and stuff that people could actually buy. I can't buy Batman's utility belt. But I can buy a Palm Pilot and use it as a database for bad guys. So can my readers, and I know that, so I write it in so they can better connect with my characters.

MORT: The comic world will always have some connection to the real world . . . but science has never been all that important in comics—except for its possibility of "far-outedness." There are more than a few popular and supposedly realistic "crime comics" today that have no more connection to real crime than the latest episode of *The Sopranos* that the writer happened to be watching while eating a pizza and playing a videogame. That's the sort of science comics best present.

Scientific fact: There really is a planet called Mars.

Comics' imaginative leap: Resurrected, stereoidinal heroes live invisibly on that planet.

MAX: More. More and more and more . . . which will probably keep writers like me in noir territory. On the other hand, when I was writing Dick Tracy, I followed Chet Gould's example and upgraded the Two-Way Wrist TV into a Wrist Computer; I did lots of technology-based stories in Dick Tracy, because the real world of crime in our era is—like everything else in modern life—bound up with technology.

**Question 6: Recent comics seem to be relying more on magic and fantasy to power superheroes than advanced science. Do you think this is because writers and artists are looking for an easier way to explain super powers?**

LEN: Absolutely. Stan Lee and Jack Kirby solved the explanation of powers issue best by creating the X-Men. They're mutants. That's why they have powers. Period.

MARK: I think it is more of a reaction against getting it "wrong" in the years to come. If the science is advanced by research, or if we gain a new insight, then a character's origin becomes an embarrassing problem. Obviously, how many people have actually gotten super powers from exposure to radiation, lightning, strange chemicals, or anything else? The entire concept of super powers is bucking the predictive curve. About the only convincing source for superpowers that we can see in modern science is in the mapping of human DNA and that was the solution I chose for creating my super people in the *Breathtaker* story. But we persist in predicting super-powered people. I suspect that eventually science will be forced to catch up with comic books.

BRETT: Making stuff up is always easier than looking stuff up!

BUDDY: Not really. I think that there's just a trend, and the audiences are buying those stories. I think the trend crosses mediums because we are seeing a lot of books and movies in these genres. I think writers today are just writing to the whims of the paying public. If tomorrow, historical biographies became the rage in comics and film, we'd see writers hunched over history books. Writers tend to move in packs.

MORT: Nope. The convoluted illogic necessary to come up with a magical universe is much harder to create than anything we can get out of a high school general science text.

MAX: The needs of the superhero myth—and it is a myth, not a SF concept—are better served by the fantasy precepts. Captain Marvel always worked better for me as a kid because "Shazam" stood for mystical, supernaturally oriented figures, a little boy literally invoking them to become a powerful man, while Superman kind of half-assedly cobbled together pseudo-science at his convenience, to justify damn near anything.

Part of this has to do with the insistence of aging comics fans to hold onto childhood concepts. Superman is a juvenile notion—a wonderful one, that speaks to the child in many of us, boys particularly—and only adults who cling to the juvenile insist on justifying Superman and his powers through "science." In this group I include not just fans but misguided writers and editors.

# Bibliography and Reading List

Along with several thousand comic books, the following books, articles, and web sites were consulted during the writing of this book.


*Aquatic Ape: Fact or Fiction?*, *The*, Souvenir Press, London, 1991.

Armbruster, John, "Manhattan Earthquake," (1/18/01) *Columbia Earth Institute News*, www.earthinstitute.columbia.edu/news.

Baker, Lois, "Procedure for Preemies Effective," (2001) *State University of New York at Buffalo Reporter*, 09/19/96, www.wings.buffalo.edu/pub lications/reporter/vol28.

Barks, Carl, *The Carl Barks Library of Walt Disney's Donald Duck Family, 1945–1974*, Another Rainbow Publishing, Scottsdale, Arizona, 1986.

———, *The Carl Barks Library of Walt Disney's Donald Duck Four Color Comics* #9–223, Another Rainbow Publishing, Scottsdale, Arizona, 1984.

———, *The Carl Barks Library of Walt Disney's Uncle Scrooge* #1–20, Another Rainbow Publishing, Scottsdale, Arizona, 1984.

———, *The Carl Barks Library of Walt Disney's Uncle Scrooge* #21–43, Another Rainbow Publishing, Scottsdale, Arizona, 1985.

———, *The Carl Barks Library of Walt Disney's Uncle Scrooge* #44–71, Another Rainbow Publishing, Scottsdale, Arizona, 1989.

———, *The Carl Barks Library of Walt Disney's Comics & Stories* #95–166, Another Rainbow Publishing, Scottsdale, Arizona, 1983.

———, *The Carl Barks Library of Walt Disney's Comics & Stories* #230–405, Another Rainbow Publishing, Scottsdale, Arizona, 1990.

———, *Uncle Scrooge McDuck, His Life & Times*, Celestial Arts, Millbrae, California, 1981.

Biehn, All, "Fluid Breathing," (2001), www.allbiehn.com/abyss/fluid breathing.

Bioscience Online, "Pheromones, What's in a Name?" (2001), www.find articles.com.

Brain, Marshall, "Question of the Day," (2001), How Stuff Works, www.howstuffworks. com/question.

Calder, Nigel, *Einstein's Universe*, The Viking Press, New York, 1979.

Claremont, Chris, et al., *From the Ashes*, Marvel Comics, New York, 1990.

———, *X-Men, Crossroads*, Marvel Comics, New York, 1998.

Claremont, Chris, and John Byrne, *The Essential X-Men Vol. 1*, Marvel Comics, New York, 1996.

———, *The Essential X-Men Vol. 2*, Marvel Comics, New York, 1997.

Claremont, Chris, and Dave Cockrum, *The Essential X-Men Vol. 3*, Marvel Comics, New York, 1998.

*The Complete EC Library, Weird Fantasy*, Russ Cochran Publisher, West Plains, Missouri, 1980.

*The Complete EC Library, Weird Science*, Russ Cochran Publisher, West Plains, Missouri, 1979.

Cooke, Robert, "Origin of All Species," (9/3/01) *Newsday*, Our Future, www.future.newsday.com.

Creation Tips, "Answers on Evolution, Creation Science, Genesis, and the Bible," (2001) *Creations Tips*, www.121.com.au/rdoolan.

Daniels, Les, *DC Comics: Sixty Years of the World's Favorite Comic Book Heroes*, Harry Abrams, Inc., New York, 1995.

———, *Marvel: Five Decades of the World's Greatest Comics*, Harry Abrams, Inc., New York, 1991.

Darling, David, *The Extraterrestrial Encyclopedia*, Three Rivers Press, New York, 2000.

Davis, Don, "Spider Facts," (09/13/00), *Explorit Science Center*, www.dcn. davis.ca.us.go.explorit.

Davis, James R., "You Only Hit That Car if You Don't Quite Stop in Time," (08/10/01), Master Strategy Group, 1992–1996, www.home. earthlink.net/~jamesdavis.

De Camp, L. Sprague, *Lost Continents*, Gnome Press, New York, 1957.

DeFalco, Tom, et al., *The Sage of the Alien Costume*, Marvel Comics, New York, 1989.

Deutsch, *The Fabric of Reality*, Penguin Books, New York & London, 1997.

DeWitt, B. S., and N. Graham, eds., *The Many-Worlds Interpretation of Quantum Mechanics*, Princeton University Press, Princeton, New Jersey, 1973.

Dispatches Human Evolution, "Water & Human Evolution," (2001), *Dispatches Human Evolution*, (Dec. 1998), www.archive.outthere.co.za/98/dec98.

Dixon, Chuck, et al., *Batman: Cataclysm*, DC Comics, New York, 1999.

Flatlow, Ira, *Rainbows, Curve Balls and Other Wonders of the Natural World Explained*, William Morrow & Co., New York, 1988.

Flying Turtle, "Super Bugs," Flying Turtle Co., (1999), www.ftexploring.com/think/superbugs.

Fox, Gardner, et al., *The Atom Archives Vol. 1.* DC Comics, New York, 2001.

———, *Mystery in Space, Pulp Fiction Library 1*, DC Comics, New York, 1999.

Gale, Bob, et al., *Batman: No Man's Land, Volume 1*, DC Comics, New York, 1999.

Galloway, Marc, "Running and the Science of Injury Prevention," (08/10/01), www.info.med.yale.edu.

Gold, Mike, ed., *The Greatest Batman Stories Ever Told*, DC Comics, New York, 1988.

———, *The Greatest Flash Stories Ever Told*, DC Comics, New York, 1991.

———, *The Greatest Golden Age Stories Ever Told*, DC Comics, New York, 1990.

———, *The Greatest Superman Stories Ever Told*, DC Comics, New York, 1987.

———, *The Greatest 1950s Stories Ever Told*, DC Comics, New York, 1990.

Gresh, Lois H., *TechnoLife 2020: A Day in the World of Tomorrow*, ECW Press, Montreal, 2001.

Gribbin, John, "Everything You Always Wanted to Know About Time Travel," (2001), www.biols.susx.ac.uk/home/John_Gribbin.

———, *Spacewarps*, Delacorte Press, New York, 1983.

Halal, William E., Michael D. Kull, and Ann Leffman, "Emerging Technologies: What's Ahead for 2001–2030," *The Futurist*, November 1998.

Hawking, Stephen, *Black Holes & Baby Universes*, Bantam, New York, 1994.

———, *A Brief History of Time*, Bantam, New York, 1988.

Hawking, Stephen, and David Filkin, *Stephen Hawking's Universe: The Cosmos Explained*, Basic Books, New York, 1998.

Hawking, Stephen, ed., *Stephen Hawking's A Brief History of Time: A Reader's Companion*, Bantam, New York, 1992.

Johnson, Clifford, "Friction," (2001), *Microsoft Encarta On-line Encyclopedia*, www.encarta.msn.com.

Jones, Dick, *Spider: The Story of a Predator and Its Prey*, Facts on File, New York, 1986.

Jurgens, Dan, et al., *Superman Transformed!* DC Comics, New York, 1998.

Kaku, Michael, *Hyperspace: A Scientific Odyssey Through Parallel Universes, Time Warps, & the Tenth Dimension*, Anchor Books, New York, March 1995.

Kaufmann III, William J., *Black Holes & Warped Spacetime*, W. W. Freeman & Co, San Francisco, 1979.

Kirby, Alex, "The Mother of All Plants," (8/4/99), *BBC News*, www.news.bbc.co.uk.

Koerner, David, and Simon LeVay, *Here Be Dragons*, Oxford University Press, New York, 2000.

KTCA Twin Cities Public Television, "The Bends," (2001), www.psb.org/ktca/newtons/11/bends.

Kurtis, Ron, "The Atom," (11/5/00), www.school-for-champions.com/science.atom.

Layzer, David, *Constructing the Universe*, Scientific American Library, New York, 1983.

Lee, Stan, and Steve Ditko, *Marvel Masterworks, The Amazing Spider-Man* #1–10, Marvel Comics, New York, 1987.

Lee, Stan, and Jack Kirby, *Marvel Masterworks, The Fantastic Four*, #1–10, Marvel Comics, New York, 1987.

———, *Marvel Masterworks, The Fantastic Four* #11–20, Marvel Comics, New York, 1988.

———, *Marvel Masterworks, The Fantastic Four* #21–30, Marvel Comics, New York, 1989.

Lee, Stan, et al., *Monster Masterworks*, Marvel Comics, New York, 1989.

Lindsay, Don, "Darwin's Gradualistic Scenarios," (2001), www.don-lindsay-archive.org.

Loeb, Jeph, et al., *Superman: No Limits*, DC Comics, New York, 2000.

Lorenz, Brenna A., "Spiders and Insects: Cousins or Strangers," About.com, (10/21/00), www.insects.about.com/library.

Lovelock, James, *Gaia: A New Look at Life on Earth*, Oxford University Press, New York, 1979.

———, *Homage to Gaia: The Life of an Independent Scientist*, Oxford University Press, New York, 2001.

Marz, Ron, et al., *Green Lantern, Baptism of Fire*, DC Comics, New York, 1999.

McMahon, Thomas A., and John Tyler Bonner, *On Size and Life*, Scientific American Library, New York, 1983.

Miller, Henry I., "Better Genes for Better Living," *The Wall Street Journal*. August 21, 1999.

Morgan, Elaine, *The Scars of Evolution*, Souvenir Press, London, 1990.

———, "Why We Are Different from Chimpanzees," (2001), www.geocities.com.

National Aquarium in Baltimore, "Fish Biology and Anatomy," (2001), www.aqua.org/education/ teachers/activities/fish anatomy.

National Institute on Drug Abuse, "The Brain's Response to Steroids," National Institutes of Health, (8/1/01), www.nida.nih.gov. MOM/ST.

Nelson, Irina, "Buckle-Up, Please!" (8/5/01), www.slcc.edu/schools/hum_scie/physics.

Nielsen, G. R., and G. B. MacCollum, "Spiders," (2001), University of Vermont, 1997, www.cir.uvm.edu.

Niven, Larry, and John Byrne, *Green Lantern: Ganthet's Tale*, DC Comics, New York, 1992.

NLECTC, "Products Database Product Information," (2001), www.nlectc.org/ASP/prodview.

Onion, Amanda, "Glowing Controversy," (2001), ABCNews.com, (8/15/01), www.abcnews.go.com/sections.science.

Palmer, Julie, "RAC confronts in utero gene therapy proposals," *Science*, October 1998, p. 272.

Pereyra, Maria, "Speed of a Bullet," (8/10/01), www.hypertextbook.com/facts.

Pilaro, Anne M., and Mercedes A. Serabian, "Safety Evaluation of Gene Therapies: Past, Present and Future," *American Society of Gene Therapy*, (1999). See www.fda.gov/cber/summaries/asgt060999ap.ppt.

Rad Journal, "Radioactive Insects Found at Nuclear Site," *Rad Journal*, 8/24/01, www.radjournal.com/news.

Ridley, Matt, *Genome: The Autobiography of a Species in 23 Chapters*, Harper Collins, New York, 1999.

Robinson, B. A., "A Brief History of the Evolution and Creation Science Conflict," Religious Tolerance.org., 1999–2001. www.religioustolerance.org.

———, "Public Beliefs about Evolution and Creation," (2001), Religious Tolerance.org, 1995–2000, www.religioustolerance.org.

Rudolph, Frederick B., and Larry V. McIntire, eds., *Biotechnology: Science, Engineering, and Ethical Challenges for the 21st Century*, Joseph Herry Press, Washington D.C., 1996, p. 12.

Schickel, Katie, "Diving Underwater," (7/5/01), Microsoft Encarta Online Encyclopedia 2001, www.encarta.msn.com.

Science Web, "The Abyss, Fluid Breathing," Starry Messenger Communications, (1996), www.scienceweb.org/movies/abyss.

Sprott, J. C., "Energetics of Walking and Running," (2001), www.sprott.physics.wisc.edu/technote.

Thinkquest, "Radiation Effects on Humans," (2001), www.library.thinkquest.org.

Thorne, Kip, *Black Holes & Time Warps*, W. W. Norton, New York, 1995.

Uslan, Michael, ed., *Mysteries in Space*, Fireside Books, New York, 1980.

Walters, Leroy, and Julie Gage Palmer, *The Ethics of Human Gene Therapy* (New York: The Oxford University Press, 1997) pp. 85–86, 91–92.

Ward, Peter D., and Donald Brownlee, *Rare Earth*, Copernicus, New York, 2000.

White, Ken, "Meco Midget Torch," (2001), www.tinmantech.com.

Whitehouse, David, "Genetically Altered Babies Born," *BBC News Online Science Service*, at: www.news.bbc.co.uk/hi/english/sci/tech/ newsid_1312000/1312708.stmI.

Wolfman, Marv, and George Perez, *Crisis on Infinite Earths*, DC Comics, New York, 2000.

Wray, Herbert, Jeffery L. Sheler, and Traci Watson, "The World after Cloning," *Online U.S. News*, (8/27/01), www.usnews.com/usnews/

Yaco, Link, and Karen Haber, *The Science of the X-Men*, BP Books, New York, 2000.

.........

# Acknowledgments

The authors would like to thank Alice Bentley, Larry Charet, Dean Koontz, Julius Schwartz, Richard Rostrom, Brother Guy Consolmagno, Lein Wein, Brett Booth, Mark Wheatley, Mort Castle, Max Allan Collins, and Buddy Scalera for their help in preparing this book. Thanks also to Lori Perkins, our agent, and to Stephen S. Power, our editor at John Wiley & Sons.

# Index

*Abyss, The*, 61
"Across the Ages," 154
*Action Comics*, xiv, xv, 1, 3, 16, 20, 33
*Action Stories*, 16
Acton, Lord, 98
Adam Strange, 147–48
adrenal glands, 29–30
*Adventure Comics*, 1
*Adventures of Bob Hope, The*, 20
aerial surveillance, 38
*Airlords of Han, The*, xiii
Alba, 31
*Alice in Wonderland*, 100
aliens, 4–8, 83–84
*All-American Comics*, 83
*All-Flash*, 116
Alliance Pharmaceuticals, 61
*All-Star Comics*, 116
*All Star Western*, 20
"All You Zombies," 157
alpha rays, 27
alternative realities, 158–59
*Amazing Fantasy*, 65, 73n
*Amazing Stories*, xii–xiii, 36
American Association for the Advancement of Science, 142
anabolic steroids, 30–32
Anderson, Poul, 78
Ant Man, 24, 25, 99–104
  square cubed law and, 101–104
Ant Man II, 100
ants, 103–104
apes, 53–55
Aquaman, 48–50. *See also* undersea heroes
aquatic ape theory (AAT), 52–56
Archangel, 130, 131
Archie, 169
*Argosy, The*, xii
Aristotle, 5
*Armageddon 2419* (Nowlan), xii–xiii
arthropods, 71
"As Never Was," 153
"Assignment in Eternity," 155
*Astounding Science Fiction*, 52
Atlantis, 48–52, 111
Atlas Comics, 20
atmosphere of a planet, 151–52
Atom, the, 20, 104–113
  suit of, 106, 109–10
  traveling through the phone line, 106–7, 110
atomic bombs, 23, 27–29, 130
atoms, 108
  solar system compared to, 111–13
  in stars, 87–96
Avengers, the, 100, 154

Barks, Carl, 161–65
Batman, 24–25, 33–45, 108, 115, 153–54, 168
  Gotham City earthquake, 43–45
  as a nonsuper superhero, 33–35
  reasons for success of, 34–35
  science of, 35–43
  utility belt of, 38–43
  villains, 35
"Battle of the Tiny Titans," 106
Beast, the, 130, 131
Bell Aerosystems, 37
"bends," 58–60, 61
Bester, Alfred, 148

*Best Western*, 47
beta rays, 27
"Better Genes for Better Living," 143n
"Beyond the Vanishing Point," 100
Bible, the, 135, 136, 140, 141
*Bill and Ted's Excellent Adventure*, 154
Binder, Otto, 26
biomechanics, 102–104
biotechnology, 142–44
Black Bull, 167
black holes, 86, 87, 155–56
Blum, Geoffrey, 162–63
Bohr, Niels, 112–13
Booth, Brett
  described, 172
  questions and answers, 175–82
*Boys from Brazil, The* (Levin), 78
Bradbury, Ray, 146
*Brave and the Bold, The*, 20, 116
breathing underwater, 51–52, 57–58
Brownlee, Donald, 11–13
Buck Rogers, xiv, xvi–xvii, 2, 36
buildering, 42–43
bullet speed, 119, 121
Burroughs, Edgar Rice, xii
Burton, Tim, 35
"By His Bootstraps," 153
Byron, John, 132

Calkins, Dick, xiii
calories, 122
Cameron, James, 61
Captain Comet, 147, 150–51, 153
Captain Future, 146
Captain Marvel, xvi, 169
carbon, 89, 90
carbon dioxide, 60, 151–52
Carson, Rachel, 151
Castle, Mort
  described, 173
  questions and answers, 175–82
Catholic Church, 5
Catwoman, 168
Centers for Disease Control, 72
Challengers of the Unknown, 20
Chandrasekhar, S., 90
Chaos Comics, 167
cheetahs, 118–19
*Children of the Atom* (Shiras), 130
Claremont, Chris, 132
Clark, Dr. Leland, 60
clones, 69, 77–81
Cockrum, Dave, 131, 132
Cohen, I. L., 140n
Collins, Max Allan
  described, 173–74
  questions and answers, 175–83
color subtraction, 97
Colossus, 131
"Colossus," 100
comets, 13
comic books
  adults and, xvii
  with all new stories, xiii–xv
  as reprint collections, xiii
  science fiction, xvi–xvii
  superhero, xv–xvii
Comic Code of American (CCA), 19–20, 68
*Comics Cavalcade*, 116
Comics Magazine Association of America, 19
comic strips, xii–xiii
*Complete Detective*, 47
computers, 43
*Congo Bill*, 20
"Conquerors of Time, The," 120
Consolmagno, Brother Guy, 17
*Contact* (Sagan), 8, 156
continuing plotlines, 21–22
Copernican Principle, 9
Copernicus, 5
cosmic rays, 21, 28–29, 120
cosmological principle, xix–xx
Crawford, Dr. Michael, 55
creationism, 136–37
  truth about, 137–42
creation science, 136
crime comics, 19–20, 116
Criminal Research Products Inc., 41
Crossgen, 167
Cummings, Ray, 100
Cyclops, 130, 131, 135
Cygnus X-1, 94

Daniels, Les, 34, 47, 65n
Dark Knight. *See* Batman
"Dark Phoenix Saga, The," 78
Darwin, Charles, 134–42
*Darwin Was Wrong* (Cohen), 140n
*Date with Judy, A*, 20
"Days of Future Past," 132

*Day the Earth Stood Still, The*, 7n
DC Comics, 1, 19–20, 24–25, 48–50, 68, 83, 105, 115, 116, 146, 147–48, 167–70
as science-oriented, 26
violence and, 34
*DC Comics: Sixty Years of the World's Favorite Comic Book Heroes* (Daniels), 34n
de Camp, L. Sprague, 51n, 52
decompression sickness, 58–60, 61
degenerate electron pressure, 89–90, 92
degenerate neutron pressure, 91, 92
Democritus of Abedera, 5
Depression, the, xi–xii
"Destination—Star," 149–50
*Detective Comics*, xiii, 20, 33–36
*Detective Stories*, 34
Deutsch, David, 158
DHA, 55
"Diamond of Deadly Dooms, The" 111
Dick Tracy, xiv
Dille, Flint, xiii
Disney, 19, 161–65
Ditko, Steve, 66, 67, 68
DNA, 80–81, 134–35
genetic therapy, 142–44
Doc Savage, xvii
Donald Duck, 161–65
Donnenfeld, Harry, xiii–xv
"Doomsday on Earth," 150–51, 153
Drake, Frank, 6–15
pseudoscience of, 7
Drake Equation, 8–11
components of, 8
doubts about, 12–15
*Driving Force: Food, Evolution and the Future, The* (Crawford), 55

earthquakes, 43–45
Earth's atmosphere, 151–53
EC Comics, 146–47, 157
"Educators Against Darwin," 140n
EGFG (enhanced green fluorescent gene), 31
Einstein, Albert, xix–xx, 38, 86, 88, 92, 125–28, 155–56
Einstein-Rosen bridge, 155–56
electromagnetic radiation, 27
electromagnetic spectrum, 27, 125
electron capture detector, 151
electrons, 112–13
Ellsworth, Whit, 34
$E=mc^2$, 88, 127
"Emerging Technologies . . . ," 143n
energy, 85–86, 95–96
entropy, 84–85
environmentalism, 151–53
Epicurus, 5
Ernest, Paul, 101
escape velocity, 18, 121
Evangelical Christians, 136–37
"Even the Heir," 157
event horizon, 93, 95, 155
Everett, Bill, 47–48, 49, 52
Everett, Hugh, III, 158–59
Everson, Brad, 142n
evolution, 6, 13, 133–137
aquatic ape theory, 52–56
case for, 133–37
creationism versus, 137–42
intelligence and, 14–15
micro-/macro-, 134
savannah hypothesis, 53–55
teaching of, 136–37
"Evolution Reward," 140n
extinction, 13, 15, 134

Fantastic Four, the, 21–29, 36, 50, 68, 99, 154
fantasy, xviii
Fawcett, 167
FDA, 143
Feldstein, Al, 146
Fiction House, 167
Finger, Bill, 34, 35, 83
fingerprint kit, 41
fish, 57–58, 60
talking to, 61–63
Flash, the, 20, 36, 83, 115–28
introducing, 115–16
problems with speed of, 117–24
revived, 116–17
speed barrier and, 125–28
Flash Gordon, 2
Fleischer, Max, 1
fluid breathing, 60–61
fluorocarbon liquids, 60–61
flying backpack, 36–38
"Flying Dutchman, The," 164–65
flying saucers, 6–7
"—For Us the Living," 159
Foster, Alan Dean, 163
Fraser-Volpe Co., 39

friction, 122
Friedrich, Gary, 129
Fundamentalism, religious, 136–42
*Futurist, The*, 143

Gable, Clark, xiv
*Gaia: A New Look at Life on Earth* (Lovelock), 152–53
Gaia theory, 151–53
Gaines, Bill, 146
Galileo, 101
Gallup poll, 141, 142
gamma bomb, 23–29, 71
"Ganthet's Tale," 84
general theory of relativity, xix–xx, 86, 92, 94
"Genetically Altered Babies Born," 143n
genetic engineering, 142–44
*Giant Batman Annual*, 39
Giant Man, 100
  square cubed law and, 101–104
*Giant-Size X-Man*, 131
"Giants of the Telescopes," 148
"Girl in the Golden Atom, The," 100
global warming, 153
God, 5–6, 137–38, 141, 142
Gold, H. L., 148
Golden Age of comics, xv–xvi, 20, 70, 83, 105, 116
Golding, William, 152
Goliath, 100
Goodman, Martin, 20–21, 24, 47, 65
Graham, Harold, 37
grandfather paradox, 156–59
graphic violence, 19–20
gravity, 14, 15–18
  stars and, 86–96
green fluorescent protein (GFP), 29–32
Green Hornet, 169
Green Lantern, 20, 35, 83–98
  black holes and, 90–96
  life and death of stars and, 86–90
  original, 83
  power source for, 85–86
  science and, 83–85
  yellow light and, 85, 96–98
Green Lantern Corps, 84, 85–86, 96, 98
Gresh, Lois H., 79n, 81n, 142n, 148n
Gribbon, John, 93n
GRS 1915+105, 94
G2 type star, 13–14
Guardians of Oa, 84, 86, 96, 98
*Gulliver's Travels* (Swift), 100

Halal, William E., 143n
Hamilton, Edmond, 26, 36, 148
Hanford Nuclear complex, 73
hard radiation, 28–29
Hardy, Alister, 52–53
Hasse, Henry, 100–101
Hatfield, Larry, 140n
Hawk Carse, 146
Hawkeye, 100
Hawking, Stephen, 94, 95
Hawkman, 20
HEAF (High Explosive Applications Facility), 42
heat and stars, 87–96
Heinlein, Robert A., 153, 157
Heisenberg, Werner, 113
Heisenberg uncertainty principle, 113
helium, 87–90
"He Who Shrank," 101
Hitchcock, Dian, 151
horror comics, 19–20, 116
"Human Fishbowl, The," 150
human genome, 79–81, 142
"Human Magnet, The," 148
Humanoid Publishers, 167
human reaction time, 120–21, 123
Human Torch, the, 48
hydrogen, 87–90
hyperspace, 95

Ice Age, 53
Iceman, 130, 131
"I Found Monstrom, the Dweller in the Black Swamp," 99
Image Comics, 167
Incredible Hulk, the, 23–25, 36, 65, 99
  gamma rays and, 25–29
  GFP and, 29–32
"Inferno," 78–79
infrared light, 18, 41
intelligence, 14–15, 53, 98
invertebrates, 71
ions, 112
iron, 90–91
Iron Man, 25
*Island of Doctor Moreau, The* (Wells), 52

Jacob, Klaus, 44
James Bond, 37, 72

jetpacks, 36–38
Johnny Quick, 116
Jones, Dick, 74, 77
*Journey into Mystery*, 25
Jupiter, 13
*Justice League of America, The*, 20

Kane, Bob, 33, 34, 35
Kerr, Roy, 155–56
Kid Flash, 120
Kirby, Jack, 21, 23–25, 66, 129
Krenkel, Roy, 146
*Kritias* (Plato), 51
Kroeyer, Karl, 165
Krypton, 11, 16–18
Kull, Michael D., 143n
Kylstra, Dr. J., 60

land bridge, 53
La Place, Pierre, 92
lasers, 38
*Last and First Men* (Stapledon), 52
latent print powder, 41
Lawrence Livermore Laboratories, 42
laws of physics, xix–xx
Lee, Ang, 25
Lee, Stan, 20–25, 65–68, 70, 74, 129, 131
Leffman, Ann, 143n
Leinster, Murray, 153
Leucippus, 5
Levin, Ira, 78
Liebowitz, Jack, xiii–xv, 20
light, 96–98, 109
  speed of, 121, 125–28, 149–50
limbic system, 30–31
Little Orphan Annie, xiv
*Living World, The* (Walker), 151
LLM-105, 42
locks, 39–40
*Lost Continents* (de Camp), 51n
Lovelock, James, 151–53
Lucas, George, 163
LX-19, 42

McIntire, Larry V., 144n
"Man in the Ant Hill, The," 100
"Many Inventions of Batman, The," 36, 38
Many-Worlds Interpretation of Quantum Mechanics, 158–59
Margulis, Lynn, 152
Mars, 151–52
*Marvel: Five Fabulous Decades of the World's Greatest Comics* (Daniels), 47n, 65n
Marvel Comics, 20–25, 47–50, 65–81, 99–100, 129, 131, 167–70
Marvel Girl, 130, 131
*Marvel Science Stories*, 47
mass of an object in motion, 127–28
M87, 94
MEC Midget Torch, 40
"Merman, The," 52
Michell, John, 92
Mighty Thor, the, 25
Milky Way, 8, 11, 94
Millennium Jet Inc., 37–38
Miller, Dr. Henry I., 143
Miller, Frank, 35
Miller, P. Schuyler, 153
miniature cameras, 39
miniature oxyacetylene torch, 40
"Mirror Master's Magic Bullet, The," 121, 123, 124
momentum, 123–24
"Moomba Is Here," 99
Moore, Wendell F., 37
*More Fun Comics*, xiii, 48, 49n
Morgan, Elaine, 54
*Motion Pictures Funnies Weekly*, 48
M stars, 14
multiverse, 158–59
mutants, 130–44
  reality of, 132–35
"Mysteries in Space," 148
*Mystery in Space*, 146–50, 154
"Mystery in the Human Thunderbolt, The," 119

naked mammals, 55
NASA, 94, 151
*National Post*, 142
natural selection, 134
*Nautilus*, USS, 164
Nazis, 48, 49
neutrons, 112
neutron star, 91–92
*New Adventure Comics*, xiii
*New Comics*, xiii
*New Fun Comics*, xiii
New Mutants, 79
Newton, Isaac, 86, 124
*New X-Men*, 131

New York City, 25
  earthquakes in, 43–45
Nicholas of Cusa, 5
Nightcrawler, 131
nitrogen, 59
Niven, Larry, 84
Nodell, Martin, 83
"No Man's Land," 43–45
Nowlan, Philip Francis, xii–xiii
nuclear fusion, 87–90

Oni Press, 167
Oppenheimer, Robert, 92, 93
*Our Army At War*, 20
oxygen, 57–60, 89, 90, 151–52
ozone layer, 151

Palmer, Julie, 143n
Palmer, Ray, 105
paradoxes, 156–59
parallel universes, 95–96, 158–59
pass keys (picklocks), 39–40
Peck, Gregory, 78
"People of the Golden Atom, The," 100
pepper spray, 40–41
perfectibility of man, 142–44
perflubron, 61
personal flying device, 36–38
pheromones, 62–63
Phoenix, the, 78
Pilaro, Anne M., 143n
"Pioneers in Science," 148
Plastic Man, 169
Plato, 5, 50–51
plenitude, 5–6, 8
pluralism, 5–6, 8
Pontifical Academy of Science, 140
Pope, the, 140
*Popular Mechanics*, 36–37
*Popular Science*, 36–37
power, 98
power source, 85–86, 96
pressure, 58–60
primary pigments, 97
primary spectrum, 97
Principle of Mediocrity, 9, 10, 12
Professor X, 130, 131
Project Ozma, 8
protons, 112
protostar, 87–88
pulps (pulp fiction), xii, xvii, 2, 47,
    145–46, 153
pulsars, 92
Punch II, 40–41
Punisher, the, 68

Quality, 167
quantum theory, 108, 113, 155, 158–59

radiation, 21–29, 66, 70–73, 91–92, 120
radio waves, 92
"Raid on the Termites, The," 101
*Rare Earth: Why Complex Life Is Uncommon in the Universe* (Ward and Brownlee),
    12–14
*Red Circle* magazine, 47
red giant star, 14, 89
Reeves, George, 1–2
"Reign of the Super-Man, The," xiv
relativity, theory of, 125–28, 141, 155–56
responsibility, 67
retconning, 29
"Return of the Ant Man, The," 100
"Return of the Super-Gorilla," 120
"Richest Man on 9 Planets, The," 148–49
Robinson, Jerry, 35
Robin the Boy Wonder, 34–35, 153–54
Romita, John, 68
Rosen, Nathan, 155
Rostrom, Robert, 17n
Roth, Werner, 129
Rudolph, Frederick B., 144n
running speed, 118–19

"Safety Evaluation of Gene Therapies . . . ,"
    143n
Sagan, Carl, 8, 156
  Drake Equation and, 9–15
saline solution, 60
savannah hypothesis, 53–55
Scalera, Buddy
  described, 173
  questions and answers, 175–82
Schwartz, Julius, 26, 36, 105, 116, 147
Schwarzschild, Karl, 92
science, xvii, xix–xx, 71, 116–17
science fiction, xiv–xviii, 2, 145–47
*Science Fiction*, xiv
"Science Snaps," 148
Scopes trial, 136
scopulae, 73

Search for Extraterrestrial Intelligence (SETI), 8–10, 15
secular humanism, 136
"Seesaw," 153
Senate hearings of the 1950s, 19, 116
senses, 123
Serabian, Mercedes A., 143n
setae, 72, 75
Shadow, the, xv
Shiras, Wilmar, 130
*Showcase*, 20, 83, 85, 104, 105, 107n, 110, 116, 117
shrinking to microscopic size, 99–101, 105–13
Shuster, Joe, xiv–xv, 3–4, 7, 16
"Sidewise in Time," 153
Siegel, Jerry, xiv–xv, 3–4, 7, 16
*Silent Spring* (Carson), 151
silicon, 90
silk, spider, 72, 76–77
Silver Age of comics, 50, 71, 105, 117, 119, 147
singularity, 92, 93, 95, 155
Sirius, 14
*Sixth Day, The*, 78, 80–81
*Sky Devils*, 47
*Skylark of Space, The* (Smith), 36
Small Rocket Lifting Device, 37
Smith, E. E., 36
smoke grenades, 41
SN1987A, 91
Snyder, Hartland, 92
solar system, 111–13
Solo Trek XFV, 37–38
"Space-Borne," 146–47
Space Race, 136
spacetime, 93, 126
space warps, 126
*Spacewarps* (Gribbon), 93n
speciation, 134
species, 134
Spectre, the, xvi
Spider-Man, 24, 65–81, 99, 108
  clones, 69, 77–81
  daring storylines, 68–69
  powers of, 66, 70–77
  radioactivity and, 70–73
  as reality-based, 66–68
  villains, 68
  webshooters, 66, 70, 75–77
spiders, 71–75
*Spiders: The Story of a Predator and Its Prey* (Jones), 74, 77
spider-sense, 75
square cubed law, 101–4
Stapledon, Olaf, 52
stars
  binary, 94
  black holes, 90–96
  habitable zone, 8–9, 13–14
  life and death of, 86–90
  mass of, 88
  in the universe, 85–86
*Star Spangled War Stories*, 20
*Star Trek*, 25, 126
*Star Wars*, 126
steroids, 30–32
Storm, 131
*Strange Adventures*, 146, 147–48, 150–51, 154, 155
*Strange Tales*, 99
Sub-Mariner, 47–50. *See also* undersea heroes
Superboy, 1
superhero comic books, xv–xvii
Superman, xv–xvi, 1–18, 25, 115, 128, 154
  as alien visitor, 4–8
  Drake Equation and, 8–15
  gravity and, 4, 15–18
  legend of, 1–3
  *Rare Earth* and, 12–15
  rewritten history of, 2–3, 16
  as super, 3–4, 16–18
supernovas, 91
super speed, 115–28
sweat, 122–23
Swift, Jonathan, 100

*Tales of Suspense*, 25, 99
*Tales to Astonish*, 25, 99, 100
Tarzan, xii
*Tarzan of the Apes* (Burroughs), xii
technobabble, 25–29, 107
*TechnoLife 2020: A Day in the World of Tomorrow* (Gresh), 79n, 81n, 142n
Teen Titans, 105
telepathy, 130–31, 133
telephone transportation, 106–7, 110
testosterone, 30–32
"Test Your Science Knowledge," 148
*Them*, 104n
  "theories," xix–xx, 138–41
  stages in proving, 138–39

thermonuclear reactions, 87–90, 93
Thomas, Roy, 129
Thorne, Kip, 156
*Thunderball*, 37
Thunderbird, 131, 132
*Timaios* (Plato), 50–51
Timely, 167
*Time Machine, The* (Wells), 153
time travel, 153–59
TNT, 42
Tobias, Phillip V., 54n, 55n
Top Cow, 167
Torpey, Frank, 47
Torres, Angelo, 146
"Trail of the False Green Lanterns," 123
transducer, 110
transgenics, 144
trichobothria, 75
*Trieste*, 164
Tuska, George, 129
TV, 1–2, 25
twins, 135

*Ultimate X-Men*, 131
ultraviolet (UV) light, 14
Uncle Scrooge, 162–65
undersea heroes, 47–63
  aquatic ancestors, 50–56
  Atlantis and, 48–52
  breathing underwater, 51–52, 57–58
  fluid breathing, 60–61
  origin of, 47–50
  pressure, 58–60
  talking to fish, 61–63
"Under the Polar Ice," 163–64
*UN-Man* (Anderson), 78
Ussher, Bishop, 135

"Vacation Time," 164
van Vogt, A. E., 153
Verhaegen, Dr. Marc, 54
Volkoff, George, 92

Walker, Wilbur, 151
walking upright, 56
wall-climbing, 73–74
*Walt Disney's Comics & Stories*, 162–65
Walters, Leroy, 143n
Wandrei, Donald, 100
Ward, Peter D., 11–13
warm-blooded creatures, 58
*War of the Worlds, The* (Wells), 4
"Was Man More Aquatic in the Past?", 52–53
Wasp, the, 100, 101
"Water and Human Evolution," 54n
webshooters, 66, 70, 75–77
Wein, Len, 131
  described, 171
  questions and answers, 171–81
*Weird Fantasy*, 146, 159
*Weird Science*, 146, 154, 157
Weisinger, Mort, 26, 48, 49, 52
Welles, Orson, 4
Wellman, Manly Wade, 26
Wells, H. G., xi, 2, 4, 52, 153
Wheatley, Mark
  described, 171–72
  questions and answers, 174–82
Wheeler, John Archibald, 93
Wheeler-Nicholson, Major Malcolm, xiii–xiv
white dwarf star, 89–90, 105, 108–9
white holes, 86, 95–96, 98
Whitehouse, Dr. David, 143n
Williamson, Al, 146, 147n
Wolverine, 131
women, 168–69
Wood, Wally, 146
*World's Finest Comics*, 1
"World's Smallest Super-Hero, The," 104–105
World War II, 19, 23, 48–49, 58
wormholes, 156

X-Men, 78–79, 129–44
  creating the, 142–44
  creationism and, 137–42
  evolution and, 133–37
  new team of, 131–32
  original members of, 130–31
  premise of, 129–31, 133
X-rays, 92, 94

Yellowjacket, 100
yellow light, 85, 96–98

"Zero Hour," 105
"Zutak! The Thing That Shouldn't Exist," 99
Zwicky, Fritz, 91

# About the Authors

**Lois H. Gresh** is the author of many popular science fiction and fantasy books for children and adults. Her titles include *The Termination Node*, *The Computers of Star Trek*, *DragonBall Z*, *Chuck Farris and the Tower of Darkness*, *Chuck Farris and the Labyrinth of Doom*, and *TechnoLife 2020* (half-fiction and half-fact). Her current project is *Chuck Farris and the Cosmic Storm*, which will be published in fall 2002. Although she doesn't write comic books—not yet, anyway—she does read plenty of them, and she's indebted to Bob Weinberg for this chance to explore the science of superheroes. Some of her favorite childhood memories involve *Tales of the Unexpected* and *House of Secrets* from DC Comics, *Batman*, *Donald Duck*, *Superboy* and *Superman*, and even a comics version of *Frankenstein*. She lives with a comics addict, her twelve-year-old son, Dan Gresh. Their home is stuffed with hundreds of comic books.

During the day, Lois serves as Creative-Technical Director of Novatek Communications, Inc. (www.novatekcom.com), in upstate New York, where she designs and creates CD- and web-based marketing and training programs.

*Visit Lois at* www.sff.net/people/lgresh

**Robert Weinberg** read his first comic book (*Blackhawk* #69, March 1953, "The Cyclone from Hell!") more than forty-five years ago and was hooked. He's collected and read comics ever since. Holding two degrees in science and mathematics, he's spent most of his life selling and writing science fiction and fantasy. For twenty years, he was chairman of the Chicago Comic Convention, the nation's second-largest comic book show.

For the past three years he's written comic book stories for Marvel and DC. For eighteen months, he chronicled the adventures of Cable, one of the best-selling X-Men titles. His latest comic, "Extinction Event," done in collaboration with artist Brett Booth, blends dinosaurs, computer technology, and high adventure in one of the most technically accurate science fiction comic books ever published.

*Visit Bob at* www.robertweinberg.net